Feminism and the Biological Body

for CRF

Feminism and
the Biological Body

LYNDA BIRKE

Rutgers University Press
New Brunswick, New Jersey

First published in the United States 2000
by Rutgers University Press, New Brunswick, New Jersey

First published in Great Britain 1999
by Edinburgh University Press,
22 George Square, Edinburgh

Library of Congress Cataloging-in-Publication Data and British Library
Cataloguing-in-Publication Data are available upon request.

ISBN 0-8135-2822-4 (cloth)
ISBN 0-8135-2823-2 (pbk.)

Printed in Great Britain

Contents

Acknowledgements

My recognition that being a scientist and being a woman do not necessarily go together in some people's minds came when I was 17; so perhaps all my work on feminism and science should begin with thanks to the young man who could not believe that I was really studying all that hard stuff (like physics), and why was I not at secretarial school?

My thinking about biology, feminism and the body also began many years ago, and developed strongly while I was doing doctoral research and throwing myself into the politics of the Women's Liberation Movement. All the women with whom I worked then in women's health groups and women and science groups thus contributed to the development of my ideas. Perhaps I should mention especially the woman (whose name I forget) who ran a workshop on vaginal self-examination at a feminist conference at the University of Sussex in the early 1970s, and who publicly demonstrated the procedure. 'Looking inside' the body took on new meaning for me in that moment.

My understanding of 'what goes on inside' bodies also owes much to the countless, equally unnamed, animal bodies I was expected to dissect in my training, or who were killed to provide tissues for student experiments. Far too many animals have died in that way.

More recently, my engagement with feminism and the body owes much to developing and teaching a course on that theme with Terry Lovell while I was in the Centre for the Study of Women and Gender at the University of Warwick. Terry, and the many students on that course, helped to expand my ideas, as have several research students, but particularly Marsha Henry and Luciana Parisi.

I am grateful, too, to other scholars who have shared their work in progress and/or ideas about feminism, biology and the body during

ACKNOWLEDGEMENTS

the period I was writing this book; particularly, Lisa Cartwright, Donna Haraway, Anne Fausto-Sterling, Ruth Hubbard, Phyllis Robinson, Kathy Marmor, Evelynn Hammonds, Mike Michael, Simon Williams, Janet Price, Margrit Shildrick and Consuelo Rivera Fuentes.

Chapter 6 began its life as an invited chapter for a book edited by Margrit Shildrick and Janet Price (1998), and published by Edinburgh University Press (although it has changed since then). I am grateful to them for the encouragement and comments that helped to develop it through its various transformations.

Reading a book-length manuscript is an onerous task, and I am especially grateful to Maureen McNeil, Consuelo Rivera Fuentes, Mike Michael, Anne Kerr and Janet Price for making the time and effort to make much-needed comments on an earlier draft. Any faults remaining are, of course, mine.

Above all, books do not get written without the patience and love of those who have to put up with the author and her seemingly endless engagement with (and some swearing at) the computer. The various members of my household have shown great forbearance (and not a little interference in the case of four-legged friends). Consuelo Rivera Fuentes patiently and with love put up with my complaints when I got stuck; more than that, she pulled me out of the mire and I finished the book.

List of Figures

Introduction

Women have long been defined by our biology. It is a familiar story; anatomy is destiny, our hormones make us mad or bad, genes determine who we are. And not surprisingly, feminists usually oppose such biological determinism, for so often it seems to fix the status quo. To be determined by biology is to surrender to limitations, to deny the possibility of change.

Feminism, however, assumes social and political change to be not only possible but desirable. Claims that, for example, men do not do the ironing because their biology makes them less able are claims that support existing gender arrangements. While I was writing an earlier draft, a book appeared, with accompanying television series, on the theme of 'why men don't iron' (Moir and Moir 1998). The author, and TV producer, believes that scientific evidence for inbuilt biological differences between women and men has been ignored – largely because of the power of feminism. Would that such power of feminism were so! On the contrary, biological claims have been abundant, but rarely stand up to close scrutiny from any source. While I rather doubt that feminists have been so powerful (I wish!), we have however been part of a necessary process of criticising such biological claims. If men do not iron, it is because they live in a society which tells them that ironing is beneath them.

The biological body has been peripheral to much feminist theory, at least partly because of that very necessary rebuttal of biological determinism. The one exception was the part played by the body in the activist work of feminist health groups (mostly in the 1970s); for to understand women's health and our relationships to the medical professions required women to grapple with medical knowledge about the (biological) body. But apart from that, the emphasis in our theory was

1

on the *social* construction of gender; the body hardly featured at all in emerging feminist theory – until recently.

Thinking about the body has, however, now become highly fashionable, reversing those earlier tendencies to ignore the body altogether (see Stacey 1997). This recent revival – or discovery – of interest in the body within feminist theory is surely welcome. It is, after all, the surface of the body which we see in the world, both our own and those of others. It is the body's surface which we can adorn or physically alter, to fit with changing cultural mores.

Yet, while this move to theorise the body is welcome, it is also puzzling to me – at least to the biologist in me. For the renewed focus seems always to end at the body's surface. If this newly theorised body has interiority, it is one that is explained predominantly through psychoanalysis (see Grosz 1994). What goes on inside the biological body remains a mystery, to be explained (if at all) only in the esoteric language of biomedicine.

Feminism is not on its own in this omission; in general, social studies of the body tend also to ignore the material inside or, at best, to relegate it to another world – the arcane world of biomedicine (by that I mean the ideas, language, and practices of the natural sciences, particularly the biological sciences, and their application in clinical contexts). While recent sociological and feminist theory has made enormously important claims about the processes of cultural inscription *on* the body, and about the cultural representation *of* the body, the body that appears in this new theory seems to be disembodied – or at the very least disembowelled. Theory, it seems, is only skin deep.

The omission to which I refer is the inside of the body as organs and physiological processes. To be sure, there is some engagement in feminist writing with genetics and with reproductive technologies: but this focuses largely on the ethics and the consequences for women of new developments in these fields. Within theory devoted *to* 'the body', there is remarkably little that enters within and considers 'the body' in terms of its own inner processes.

Here, then, I want to look 'inside' the body, particularly at the ways in which we might come to understand 'what goes on' physiologically. In that sense, I am concerned not only with the physiology of human bodies, but also with how we in Western culture learn to perceive that physiology. Curiously, students I have taught in classes on feminism and the body seem to drop much of their critical stance when asked to think about images of the body's insides. Deconstructing images or texts has become second nature to students of women's studies; they will happily do it with regard to most forms of cultural production, including some parts of science (new developments in genetics, for instance). Yet our bodies' insides seem so often to be stuck in a language of certain facts – here is a Fallopian tube, there an ovary. Only when those inside processes are externalised, brought out by technologies such as whole body scanners, do they seem to become exposed to view.

My wanting to 'look inside' the biological body was, then, born of my frustration at the gap between feminist cultural analysis and my own background as a biologist. As I worked on this book, I kept coming back to my own memories of that scientific training. How did I learn to do the experiments? How did I learn to label body parts? What assumptions underpinned that training? And – more importantly for this book – how did I re-learn as my critical feminist self re-membered? Throughout, I have interspersed some of these memories and reflections; in a sense, then, this book has become a soliloquy on my travels through a particular form of (Western) scientific training and stories about how bodies work.

But to contextualise that voyage, I begin the book by looking at the background to feminist challenges to science (and particularly biology), and at the problems those challenges provoke. Among the questions I need to ask, for example, is what is it that we mean by 'biology'? From there, in Chapter 2, I outline some significant ideas about the body emerging from feminist theory, and at feminist critiques of how the body becomes gendered in the narratives of the biological sciences.

These explorations set the scene for my later forays into the physiological body. I begin these with anatomy as a source of knowledge about the body, focusing in Chapters 3 and 4 on how bodily insides are represented through diagrams and how these then influence how people might think about their insides. I then move on in Chapter 5 to think about some basic principles of physiology; in particular, I explore the social contexts out of which influential ideas in physiology – such as homeostasis – emerged.

From there, I consider a 'case study' of a specific part of the body – the heart (Chapter 6). Here, I draw on the themes of inner space and control outlined in previous chapters and illustrate them in relation to culturally salient ideas about the heart, and how these might influence medical practices.

Chapter 7 picks up again feminist critiques of assumptions in biological science, and asks whether we might seek (or find) other narratives. A substantial part of feminist critique has noted the power of reductionism within science – that is, the belief that phenomena can be reduced to simpler or smaller ones (like the belief that everything we do is caused by our genes). So, one response to that is to seek less reductionist narratives; some of these we may draw from outside the narrow domain of 'the natural sciences' as they are formally taught in Western universities – the various practices and narratives of alternative medicine is one such source. Others, however, can be found even within what is conventionally called science, and it is these that I explore here in order to see what they might (or might not) offer us.

Now I should emphasise at this point that by referring to 'other narratives', I do not wish to imply that these are simply alternatives, of equal worth. Rather, I argue that certain theories or ways of conceptualising biological processes are better, more truthful, in the sense that they describe more accurately (or in less reductionist ways) how living organisms work. They may also be better in feminist terms, in the sense that they support reductionism less strongly. I should also emphasise that, while I use the phrase 'other stories' to indicate these alternative frameworks, I do believe that they provide some sort of account of nature 'out there'. Our understanding of biological processes may indeed be mediated through culture, as many theorists have insisted; but it does have some relationship to 'nature', however complexly understood or negotiated.

Finally, in Chapter 8, I bring together some of the themes I have identified in my journey through the biological body, and consider some of their implications. In doing so, I make connections back to the feminist theorising of the body outlined at the beginning of the book; I also speculate on further cultural and political connections. Throughout, this book emphasises my desire *not* to lose the 'biological body' in feminist analyses, a theme I stress in the final conclusions.

There are many topics or approaches I could have chosen – the selection here is arbitrary. What I do not do is to focus on reproduction – though that would be an obvious point at which to engage feminist theory and the biological body. Reproduction (and, relatedly, genetics and reproductive technologies) have been central to much feminist engagement with biology. But I wanted to move away from these familiar topics, to try to think about 'the biological body' and its gendered constructions through other themes.

My focus is Western science, because that is what I know best. There are, to be sure, other sciences and other ways of conceptualising the 'natural world'. But modern science and technology have enormous power, not only to name the world but also to exert influence over it, for good or ill.[1] That science assumes the mantle of a universal truth, from which we must infer that descriptions of how bodies work are culturally neutral. They are not, of course; those descriptions are deeply embedded in the history of what we now understand as the natural sciences, as these emerged at a time of Western expansion. So, the images that I describe in, for example, scientific diagrams of the body are images that reflect the culture in which they have developed. They thus carry the hidden imprint of a colonial history (see Anderson 1992; Arnold 1993; Schiebinger 1993). Through these various representations, the scientific story of 'the (universal) human body' is told, throughout the world, in many cultures. And whose body is it typically found in the diagrams and texts?

In each chapter of this book, my concern is to examine how we (in Western culture) might think about some part or aspect of the body; what kind of narratives structure the scientific tales told about our bodies? While acknowledging the deep cultural assumptions pervading scientific representations of the body, I have focused on reading them in the context of the Western culture in which I live and in which I trained to be a scientist. The kinds of question that underlie my writing are: How do the scientific images of how the body works relate to/structure people's understanding of the 'real body' – to what extent do we all take on board the language of biomedicine, with its reductionist assumptions? To what extent are assumptions of gender/race/class/sexuality or disability built into these representations of bodily insides? And what implications do these questions have for feminist work and politics?

1

Ironing out the Differences? Feminism and Biology

> ...dashing away with the smoothing iron,
> she stole my heart away...
> [traditional folksong]

I loathe ironing; I usually avoid it. By contrast, several close male friends seem almost to enjoy it. I mention this because ironing happens to be part of some recent claims that 'women's work' is what women do because of some underlying biology (Moir and Moir 1998). If men do not do it, the reasoning goes, this is also because of their biology.

The gender difference is familiar; what would that line from the song above look like with another pronoun? Even without the pronoun 'she' we would read the lyric as being about a woman (and probably about a servant, doing the ironing for the household). Stealing the narrator's heart, too, is likely to be a feminine trait, for the metaphor of the heart as the seat of emotion and love is a powerful one, with overtones of gender.

We can hope that the narrator does not literally lose his heart, even in these days of heart transplantation. We know, of course, that the reference is metaphorical, to the emotions. Yet those narratives of the heart of emotion and the heart as mechanism coexist in Western culture at the end of the twentieth century. Both inform how we, in this time, might experience our bodies.

If we want to describe our bodily insides, we may borrow the metaphors buried in the arcane language of science; but they sit alongside rich cultural histories of other ways of thinking about the inside of the

6

body. The heart, for example, may be 'just a pump' to the transplant surgeon, but has a history of being a repository of emotion in Western culture. We can call upon the language of pumps, we can visualise working parts of our anatomy with technology; yet we might still, in other contexts, draw upon imagery of hearts pierced by arrows of love. However fondly we might etch outlines of hearts onto trees, it is not these that would be displayed by technologies of cardiac monitors.

The focus of this book is the scientific accounts of what goes on inside bodies, and the cultural contexts of those stories. Whatever narratives science offers, other stories also circulate, and can inform how any of us might think about our bodily insides. My starting point is my long-standing engagement with gender and science. Like other feminist biologists, the focus of much of my work has been on the ways in which our understanding of the natural world is gendered. We have, for example, criticised the assumptions behind claims that gender difference in human behaviour is rooted in biology – be that ironing or becoming chair(man) of the company board. We have also critically examined narratives of the gendered body, which so often seem to describe women's bodies as somehow deficient. Even the heart, which at first sight may seem genderless in the scientific narratives, does not escape a gendered reading, as I will note in later chapters.

What is at issue for feminist critiques is how assumptions about gender – about masculinity and femininity – are read onto nature, including the insides of our bodies. From studies of the behaviour of primates (Haraway 1989) to the behaviour of molecules or bacteria (Spanier 1995), we can find gendered assumptions written into accounts of 'how nature works'. Gender, a complex set of social and cultural practices, lurks behind many textbook narratives – so making it much easier to interpret gender *as* the product of nature. It is precisely that double move that has been the focus of a great deal of feminist analyses of the natural sciences.

This chapter is concerned with feminist critiques of science, particularly biology. My own work in this area comes out of my own experiences as a scientist, and as a feminist – and from the contradictions that those two sets of experiences brought. From there, I move on to outline some of the major themes in feminist analyses of biology, beginning with the insights of the women's health movement in the 1970s and moving on to the more recent critiques of science as knowledge. But underlying all this is one important question; what do we mean by the term 'biology'? I address this question at the end of the chapter.

My thinking about the biological body is rooted, then, in feminist analyses of the 'natural' sciences. It also draws upon recent work on the body in sociology and cultural studies, which documents the various ways in which (Western) bodies can be controlled or changed. Another area of scholarship on which I draw is the sociology of scientific knowledge, concerned with the social construction and negotiation of knowledge in science. What counts as scientific knowledge, as the 'facts', depends on who counts it as such and in what context. 'Nature' is only one player in the game of knowledge creation that we call science. As we shall see in later chapters, how we have come to understand the inner workings of the body is deeply dependent upon particular configurations of people and events.

This project is also rooted in my own experiences; I cannot reflect upon science and its knowledge without also thinking about how I, as a budding scientist, came to that knowledge. How, for example, did I learn to interpret what I saw in the experiment, or down the microscope? What did that process of 'learning to see' do to me, or for me, in my learning science? And – equally important – how did my engagement with feminism enable me to question the scientific learning, to 'read against the grain' of scientific texts?

Nor can I reflect upon the biological body without remembering how powerfully biomedicine determines our experiences/understandings of bodily insides. I have seen (too) many X-rays of my ribcage as a child (done because my grandfather died of tuberculosis); how could I fail to learn what ribs looked like? Today, ultrasound imaging is commonplace – in illness, and particularly in pregnancy – along with other technological means of visualising our insides. These transform our experiences of health and illness; they also transform our relationship to our insides. If the technology generates a picture of something that 'should not be there' – a shadow, a lump – then we immediately move from being healthy to being potentially sick, whatever we *feel* like. How do experiences like these structure our understandings of the inner body, as its secret recesses are probed and exposed?

My own thinking and writing about the biological body have, above all, been influenced by my own background *as* a biologist. I decided to 'become a scientist' when I was very young – whatever that choice meant to my young mind. At the time, I did not know that girls were not meant to do science; neither my parents nor my school (a girls' school) told me about that. Only later did I crash headlong into the wall of suspicion of girls who did science, and which I suspect (with hindsight) helped to push me away from the physical sciences I then preferred and towards biology.

8

In addition, I have always loved natural history, and being around animals; biology allowed me to indulge that love. It also meant doing things that I disliked intensely, such as cutting up dead animals (I have written about the experience of dissection elsewhere: see Birke 1995). But it also reinforced the lessons learned from my childhood plastic models (the 'Visible Woman' was one); swallowing my disgust and the smell I traced the way the cranial nerves radiate out from the central nervous system of the dogfish or rat, or examined the way that the organs fit together.

Later, my fascination for how bodies work (somewhat) overcame my squeamishness, to the point where I wanted to study more physiology, particularly focusing on the nervous system. I went on to do research that drew on my knowledge of physiology. That background, and what it taught me,[2] underlie the themes of this book.

I was an undergraduate in the late 1960s, at a time of student unrest and political protest. No one, not even a science student, could fail to notice. I became politicised, and began to make links to the emerging radical science movement, and to the women's liberation movement. Among other things, these political activities put me in touch with women's health groups, with their practical concerns about women's bodies. Thus, I became engaged with struggles to protect women's health and reproductive rights, and with women learning to challenge medical power. Among other things, women's health groups at the time often practiced self-examination. Looking into our inner spaces (at least of our vaginas) with a speculum was (and still is) a radical challenge to medical power (see Boston Women's Health Collective 1973).

That political engagement also pushed me towards criticism of the science I had been studying. I have been on that road ever since. Given the power of science, it is imperative for feminists to critique it, to analyse the multiple ways in which modern science has contributed to sexism, racism, or other kinds of inequalities. We need to understand also how assumptions founded on such inequalities become built into the ideas of science. Stereotypes of gender run deep in the history of ideas about the body, for example.

So too does a whole set of assumptions about the appropriateness of using animal bodies as 'models' for the human body. To draw upon the physiological models of my training meant having to deal with the recognition that that knowledge required the dismemberment of many such animal bodies. The experiments I read about involved the 'sacrifice' of untold numbers of sentient animals, so that humankind could learn a little more, perhaps, about how animal bodies work. My fascination for finding out about how they work has become tempered by my concerns about the brutal methods by which *we* come to find out. Studying physiology required that I put that concern out of my mind for a while.

The predominant metaphors for the biological body draw inevitably from biomedicine. Here, we can find the facts about how the body works – or so we might believe. The more I engaged with feminist politics and critiques, the more I came to question the idea of unadulterated 'facts', and the power of the knowledge and narratives of science. How does it acquire its cultural authority? What is it about *scientific* knowledge and knowledge claims that give them that power to persuade (see Latour 1987)?

Feminist and other work in science studies pays much attention to the discourses of science; what is it about that language that can be so authoritative? Whether in written text, formal presentations at conferences, or in personal chat in the laboratory, scientists must use the rhetoric of persuasion (see Knorr-Cetina 1983; Gross 1990; Myers 1990). Their findings will count for nought if no one else can be persuaded of their truth. Analysing biological writing, Greg Myers (1990) notes how easy it is for the non-scientist to lose sight of the fact that scientific articles are texts: they are rhetorical. 'That we do not see the armies of the other interpretive options – the losing views of phenomena – is only because in this battle, the losing army is immediately buried. We see only the shining armor of the facts that remain', he argues (ibid., p. 259).

Scientific texts, like any others, draw upon whatever narratives are culturally available; powerful metaphors and gendered fables are perhaps to be expected. My growing feminist consciousness made me more aware of those losing armies of interpretation, the 'other stories' that might be told about how nature works. It was quite a shock: the facts of science that I had hitherto taken for granted, more or less, suddenly were called into question. From then on, I would not be content with 'merely' learning the outcome of experiments, the hypotheses and data, the theories and findings; I would also want to know what assumptions lay beneath, to understand the rhetoric and trace the metaphors.

So to explore what lies beneath the body surface means exploring some of the assumptions and language built into the scientific stories. To follow these various threads means also that I must weave between feminism, the narratives of biomedicine, of sociology and history of science, as well as my own personal experiences of each of those.

Shifting metaphors, I might borrow Emily Martin's 'rhizomatic' approach (based on Deleuze and Guattari 1987), tracing the cultural connections between ideas of the internal body as they spread through the soil of our culture (Martin 1996, p. 102).

Martin notes the still-prevalent mechanistic and reductionist rhetoric of science (to which many scientists still adhere, even while paying lip service to other interpretations). Yet, she points out, there is also a wider cultural shift towards interpreting the body in non-mechanistic ways – the growing fascination with 'alternative medicine' illustrates this trend. At present, scientists who insist on thinking of biological processes in such ways are in a minority. But, given shifting percep-tions within the culture of which scientists are part, then that must change, Martin argues (ibid., p. 106). Much of my focus here is on the texts and practices of biomedicine, which are mostly mechanistic; later in the book I will (re)examine moves away from mechanism within science, to ask how we might use such ideas in thinking about the body.

While I will be looking critically at scientific texts and their narra-tives, my reexamination of them also reminded me of why I have always found science fascinating. Masculine metaphors of engines, militarism, and so forth notwithstanding, I still find it wonderful to read about how bodies work. Living organisms have exploited all parts of the globe, all kinds of different environments. How their bodies have evolved to become adapted to such enormous differences in temper-ature, humidity, salinity and so on is endlessly fascinating to me. That fascination remains central to my thinking about science, even while I play the part of feminist critic. It is partly for that reason that I seek ways to bring biology into feminist theory rather than simply rejecting it as inevitably determinist.

FEMINISM AND BIOLOGY: OUR BODILY SELVES?

Our bodies, ourselves – an evocative phrase. In 1970s Western feminism, bodies were acknowledged in the action of women's health groups, and in campaigns for reproductive rights. A central concern of the self-help health groups of the time was women's empowerment, through making accessible to women medical knowledge about 'how the body works'. The approach is epitomised by the publication of the Boston Women's Health Collective's well-known book, *Our Bodies, Ourselves* (1973), as well as a number of other self-help books (see for example, Shodhini 1997).

These ground-breaking books sought to reclaim women's health – at least partly – from medical control. They did so, first, by listening to what women themselves said about their bodies and experiences of bodily processes, and second by linking these experiences to knowledge of how bodies work in the physiological sense. The aim was to empower women, to allow them to take some control over their own health and bodies. Among other things, the lavish use of drawings and photographs in *Our Bodies, Ourselves* was intended to be celebratory – of women and of our bodies and body parts.

Yet that concern to understand the inside of the body was limited. First, it was limited by its inevitable focus on reproduction. Women have, of course, struggled for greater control over their own reproduction – and continue to do so. But women have also long been subject to medical ideologies that construct us as little more than wombs on legs, which a feminist focus on reproduction does not always challenge. Indeed, I would hope that the 'self' evoked by the familiar phrase 'our bodies, ourselves' *is* more than a set of reproductive organs.

Second, the various health books tended to describe the anatomy and physiology of the body, presenting simplified versions of what appeared in medical textbooks. That it *was* explained simply was undoubtedly important, enabling many more women to gain knowledge about their bodies. But, as one member of the Boston Collective later lamented, their text uncritically repeated the language of the biomedical accounts and their assumptions of uncontested facts (Bell 1994). The language of menstruation and menopause as deficiencies – so common in medical texts (see Martin 1987) – reappeared in the feminist health books.

Third, the growth of women's health groups during the 1970s was running in parallel to the growth of women's studies and the development of feminist theory. At the time, a central theme was to insist on analysing gender as socially constructed and apart from the biological distinction of sex. Not only was 'sex' seen largely as a biological bedrock, but biology was explicitly contrasted to social construction. Emerging feminist theory focused on social and cultural determinants of gender, leaving the body and its biology to the more practical domain of women's health groups. Indeed, the very focus in women's health books on control *over* the body helped to reinforce the separation of biological body from social self.

These distinctions continue, particularly in relation to health. A recent publication by the World Health Organisation, for example (PAHO/WHO 1997; see pp. 56–7) illustrates its main themes of gender and health using a diagram to show health needs. The diagram

12

recognises cultural contexts, showing in outer circles 'gender roles', 'access and control of resources', and 'aptitudes and skills', for example. But at the centre lies another circle, somehow sacrosanct: this is biology, the biology of 'sex', apparently untouched by other contexts.

One, almost inevitable, consequence of this widespread division between 'biology' and the 'social' has been that whatever was assigned to the category 'biology' was ignored or seen as inaccessible to cultural analysis. So, other kinds of animals, plants, and the interior workings of the human body, are typically classified as biology. For example, feminist theorists of the 1970s–80s were concerned to analyse gender as separate from 'sex': sex belonged to biology, and was largely unanalysed.

Part of the reason for this gap is because of the long-standing separation of what counts as biology from other fields of scholarship. So, while new theories of the body emerge in sociology and women's studies, the 'biological body' remains forlornly outside. My own history as a biologist and feminist bears witness to that separation: it is hard work indeed to bring those two facets together.

Science was not, however, left out of the burgeoning feminist critiques of the disciplines during the 1970s. Not content with merely pointing to the dearth of women in scientific occupations (important though that issue is), feminists also sought a more radical challenge. And what made these analyses and activism around science radical was precisely the way they questioned the content and practice of science. Science is important not only in its exclusion of women (and other groups), but also because of its power to describe and name nature.

Feminist science groups of the 1970s drew upon insights gained from other political issues; many of us involved, for example, were also engaged in the radical science movement, which had grown up in the late 1960s from the awareness of the part played by science in war and genocide, and global environmental damage. Unlike the optimism towards science of the 1930s, those who grew up after 1945 were only too well aware of the potential of science to destroy. Not surprisingly, then, feminist critiques of the 1970s focused intensively on the destructive power of science – whether that be through power over women, over non-white people, over non-heterosexuals, over the environment, over other creatures (see, for example, Brighton Women and Science Group 1980, who subtitled their book 'The power of science over women's lives').

After 1980, however, there was a perceptible (if initially slow) growth in feminist scholarship focusing on science and technology. Feminist scholarship began to move into new areas from its initial

13

preoccupation with biology and medicine, now including work on physics and much more on technology, as well as the history and philosophy of science.

Yet the focus on biological ideas remains, for it is such ideas that so directly have framed theories about the 'inevitability' of existing gender roles. And, as Hilary Rose has emphasised, it was no accident that biological claims about women's role should intensify just at the time that women were organising collectively in political struggle (Rose 1997) – just as biological claims about homosexuality are now increasing in the wake of greater lesbian and gay visibility.

The earliest feminist critiques of science focused on issues of bio-medicine (Brighton Women and Science Group 1980; Hubbard *et al.* 1982), asking questions about, for example, evolutionary theory, brain structure, and hormonal theories of lesbianism. These essays dealt with ways in which human behaviour was allegedly determined by underlying biology, and on the extreme reductionism of such arguments (that is, that complex social behaviour is explained, sometimes exclusively, by reference to something inside the body – a molecule, for example). The essays noted how biologically determinist arguments repeatedly ignore the social and cultural context in which gender, race and other social markers are created. Again and again, what we found was unwarranted assumptions, sloppy data, and flawed conclusions: so much for objectivity. Science, the queen of rationality and objectivity, was turning out to be rather inclined towards partiality.

Not surprisingly, then, one major area of feminist critique of science has focused on objectivity. Science, the story goes, epitomises the pursuit of objective truth, the exercise of supreme rationality; it seeks to tell the true story about the world out there. But we can challenge this tale, insisting that what scientists produce is culturally and socially embedded. That is not to say that the knowledge of science is pure social construction; for many of us taking part in those challenges, what needs to be understood is *how* our understandings, and their social embeddedness, relate to the world 'out there' that we seek to describe, and vice versa. As feminist philosopher of science Sandra Harding has argued (1991), no one can completely distance themselves from their social and cultural context; thus, the objectivity attained must become an objectivity rooted only in the experiences of particular people (those who get to do science). It can thus only be a weak objectivity; if science were more fully to represent the wider society (that is, if more kinds of people were able to contribute to it substantively), then – and only then – might it approach the strong objectivity which it currently claims.

Meanwhile, denial of human agency in science, and insistence on what Donna Haraway has called the 'god's eye view' of the world (Haraway 1991c), predominate. Things get done in scientific reports: no one, it seems, actually *does* them. Like other scientists, I had to learn to write like that, to remove myself from active participation, to emphasise reductionist conclusions and omit the messier details.[3] And that god's eye view was protective: it at least meant that I did not overtly have to admit what I well knew – that my feminist interests were involved in my choice of research topics (see Birke 1995). Science is never disinterested; it is just written up as though it is.

That denial of human agency contributes both to the benefits of science and to its destructive power. The benefits arise from the way that 'objectivity' facilitates some measure of prediction and control (and we have all benefited in various ways from developments in science and technology based on that predictability). Yet is also the distancing stance which helps to generate the inhumanity and injustice which feminist critics have identified. It is not only women who have suffered from the consequences of the 'god's eye view': so have certain other people and other species. So, one very important strand of feminist critique accordingly concerns itself with the wider destructive power of science – documenting, for example, how modern science has contributed to environmental destruction (Merchant 1982; Gaard 1993; Plumwood 1993; Shiva 1995), and to the abuse of animals (Donovan 1990; Adams 1994; Birke 1994).

To take an example of how the distancing stance of objectivity contributes to inhumanity, Karen Messing and Donna Mergler (1995) examine how occupational health research is conducted in ways that ignore or play down the interests of workers and/or women, frequently rendering it inhumane; the research, they point out, is deeply imbued with particular assumptions about gender, race and class. What matters to researchers is objectivity; so, carefully controlled trials matter more than the ethics of exposing a so-called 'control' group of people to danger without their knowledge/consent. What perpetuates that inhumanity is precisely the 'god's eye' stance, in which objectivity matters more than people (or other sentient beings).

The distancing ignores context – not only the social context of the observer (and the sociocultural assumptions she or he brings) but also the context of whatever it is under study. Contexts disappear in the reductionist focus on one or two variables to be investigated. That is one reason why feminists have insisted on critiquing reductionism: for not only does reductionism imply a kind of violence towards nature in its assumption that nature can best (or only) be understood by analysing component parts, but reductionism can lead to 'magic bullet' answers. So, rather than try to understand the complexity of a problem, scientists or doctors sometimes turn to surgery, drugs, or genetic solutions. As Hubbard and Wald have noted, concerning claims for a 'gene for learning disability', genetics can come in handy in letting educational systems off the hook (Hubbard and Wald 1993).

Alongside these critiques of the assumptions underpinning scientific research, more detailed feminist analyses of biology appeared (for example, Bleier 1984; Fausto-Sterling 1985; Birke 1986; Hubbard 1990; Rosser 1992). What these works shared was a concern to uncover what Anne Fausto-Sterling (ibid.) called 'the myths of gender' in scientific accounts; how, for example, did scientific narratives account for gender? What assumptions did they make? How much did cultural stereotypes of appropriate gender behaviour find their way into the supposedly objective textbooks of science? Unsurprisingly, perhaps, scientific accounts were not averse to reiterating age-old stereotypes; from heroic sperm and promiscuously masculine genes to the coy, sedentary egg, nature is deeply gendered in scientific narratives (Biology and Gender Study Group 1989; Martin 1989).

Other critics looked at a range of topics within biology. Studies of animal behaviour, and the extrapolation to humans from what scientists claimed about certain species of non-human animals, were a rich source of feminist criticism. Perhaps inevitably, the intersection of gender and race with scientific ideas is clearest in relation to primates – our closest relatives, onto whom we project all manner of assumptions (see Haraway 1989).

Animal behaviour and physiology are the areas in which I was trained, in which I specialised. So it is not surprising that I paid these particular attention with the lens of feminist criticism. How scientists describe the behaviour of non-human animals is undoubtedly a place to seek sexist and racist assumptions about humanity, as Haraway reminds us. But I should also add that it is among the scientists who study animal behaviour that I have found the most awareness of the role of the researcher, and her/his potential to influence the animals being observed (see Birke 1995); in the study of animal behaviour, if in few other areas of science, there is some criticism of the 'god's eye trick' and an understanding that we cannot always distance ourselves so completely.

Other areas for feminist attention have included ecology (Gross and Averill 1983; Plumwood 1993); immunology (Haraway 1991b; Martin

1994); endocrinology, and the assumption that hormones are gendered (Oudshoorn 1994; Wijngaard 1997); evolutionary theory and molecular biology (Keller 1992; Masters 1995); bioethics, including new reproductive technologies and genetic engineering (for example, Hubbard 1990); the relation between concepts of humans and other animals (Birke 1994); and the body and its developmental processes (Bleier 1984; Birke 1986; Fausto-Sterling 1992).

A great deal of feminist attention to science, however, has focused on reproductive technologies, and on the 'new genetics' (including genetic engineering). Certainly, developments in biotechnology pose many challenges to women globally. Vandana Shiva (1995) argues, for example, that genetic engineering can create new organisms unsuitable for specific environments; she further raises the question of ownership of knowledge. Who 'owns' the knowledge of how plants in the rain forests can be used in healing? We can be sure that if battles for ownership ensue, it will not be the indigenous peoples who win, she argues. The shift towards seeing 'the' gene as the explanation of everything is, critics insist, a dangerous one; to return to Hubbard's point, it lets institutions off the hook and invokes the spectre of eugenics. A substantial part of feminist work on the consequences of the multimillion dollar efforts to 'map the human genome', for instance, has centred on the collection of DNA from threatened human populations (the Human Genome Diversity Project). To advocates of gene mapping, it matters that they analyse DNA from various different human populations, to help them to understand something about patterns of movement and human origins.

Jetting in with a hypodermic and jetting out with a blood sample is, however, a new form of exploitation by the affluent north, and has accordingly encountered protest; neither the blood sample nor the resultant DNA map will come back to the people from whom they were taken (see Shiva 1995). Why not, ask protesters, spend that money on helping to preserve the populations in question, and their cultural heritages, rather than spending it on taking their blood? People, their bodily integrity and cultural histories, become redundant in the zeal to map their genes.

Apart from overt political struggles over the introduction of new gene products, critics also point to the impoverished account that genetic determinism offers. To many spokespeople for the gene mapping initiatives, it is possible to locate a 'gene for' something or other. That something may be a specific disease – and, indeed, some genes have been located which appear to be implicated in specific diseases (cystic fibrosis, sickle-cell anaemia, and Huntington's disease, for example).[4] But all too easily the same apologists speak of genes 'for' complex behaviour, such as sexual orientation (Dean Hamer's claim that his team had found a gene coding for male homosexual orientation is one example: Hamer *et al.* 1993).

Such claims further illustrate how context so easily disappears. Like other instances of biological reductionism, this example can be analysed in terms of the quality of the science, as feminist critics have indeed done and must continue to do (see Fausto-Sterling 1992). Another part of the critique has focused on the underlying assumption that having, or not having, a particular gene is what really matters in determining who you like to have sex with.

There are rather too many examples of such determinism. The juggernaut of modern genetics continues to generate them (even while geneticists sometimes try to disavow the extremes of determinism[5]). The context of a person's life (or any other organism's) seems to disappear in these accounts, as does the context in which particular genes may become effective. So too do the complex processes by which fertilised eggs become embryos become foetuses, or by which foetuses are able to develop into humans with being. I will deal with some of these issues later in this book, in relation to thinking about the biological body and agency. It is precisely those issues of how whole organisms – including whole bodies – come into being, that are omitted from genetic determinism. They are also omitted from social or feminist theories of the body.

In the flurry of scientific activity around genes, the organism as a whole entity seems almost to disappear. The preoccupation of feminist critics with biotechnology further facilitates this loss, as genes (or perhaps the social/ethical consequences of genetic research) become the focus even of feminist inquiry. Whole organisms, including our bodily selves, appear irrelevant in the world of genes: indeed, some enthusiasts for the genes'R'us story do think bodies are secondary – they are for instance merely lumbering robots at the bidding of the immortal genes in Richard Dawkins's book extolling 'selfish genes' (1976).

For now, however, we should note that there does seem to be a sea-change, even among biotechnologists. Even one of the leading journals, *Nature Biotechnology* has been publishing commentary on 'epistemological questions' (end of 1997). This, I must stress, is a breakthrough: the notion that a scientific journal might ask about epistemology at all shocked me into disbelief when I first saw it. Genetics is beginning to shift, to take account of the contexts in which genes operate, including the cells and tissues around them. As Evelyn Fox Keller wryly puts it: 'A funny thing happened on the way to the holy grail' (Keller 1995; p. 22); geneticists, that is, found themselves increasingly unable to assimilate their findings to their reductionist framework. Now, Keller argues, 'molecular biologists, it appears, have "discovered the organism"' (ibid., p. 29) – the context(s) in which genes operate.

The extent to which feminist analyses of biology – and of cultural representations of the biological body – might draw on these 'alternative frameworks' is a theme of the last part of this book. Like other feminist biologists, I want to reject biological determinism and to seek what I see as more contextual ways of describing/explaining biological processes.[6] My rejection is, of course, partly political – determinism limits possibilities of change – but it is also because the reductionist accounts offer a very limited understanding of how living things work. Simply saying that an animal does something because of a particular hormone, say, ignores all the other influences in the animal's life. Might we, then, seek other ways of describing nature, which do not ignore context?

A large part of the political work in challenging science has, however, been to criticise the methods and assumptions underpinning particular claims. Like other feminist biologists, I have at times got quite fed up with having to deal, again and again, with claims that scientists have discovered a gene for this, a hormone for that, or an evolutionary pattern to show how women should love housework. Biology, of course, is the culprit, and so I find myself yet again the biologist critiquing biology. The task is like the Aegean stable; however hard you work, the shit keeps piling up.

Yet the task must be done, however unpalatable, and we have necessarily to go on shovelling. At one level, this requires analysing and critiquing the methods by which the science is done: Are the sample sizes right? Are they appropriate? Are the statistics adequate? Are the conclusions legitimate? This tactic assumes, of course, that the science that generates the deterministic ideas we challenge is bad science that could be improved. Whatever the problems with that assumption (it assumes that what is wrong is simply method, rather than the location of science in a particular kind of society; see Longino 1990), it has remained a necessary tactic.

Feminists have – rightly – been highly critical of biological determinism; that is, we have rejected notions that social distinctions (of gender, race, ethnicity or class for example) are caused by underlying biological distinctions. Having played my part in the necessary criticism of such notions, I must confess to some qualms. There are underlying problems with our critiques of biological determinism, that we have failed adequately to address. One of these is that, in questioning biological determinism, we have insisted upon social construction – of gender, of sexuality, of the inequalities of race. A corollary of this position is that we thereby perpetuated the distinction between what counts as sociocultural (human) and what counts as biology. We can then ignore the latter, or leave it to the arcane interests of biologists.

The narratives of biology, however, help to define what counts as normal function and what counts as pathology. They help to define the 'normal body' and to delineate 'deviant bodies'. Our bodies – our selves – thus become categorised through the language of biomedicine. That is reason enough to take biology more seriously. If our bodies are indeed our selves, as the title of the women's health handbook suggests, then we need to pay more attention to how they work and to how we understand their workings.

WHAT COUNTS AS BIOLOGY?

What does the term 'biology' connote? It can mean a particular discipline, part of the natural sciences: 'biology' implies the study of living organisms – plants, animals, and microorganisms – and their processes. The word can also be synonymous with those processes, as in 'human biology'. The term 'biology' all too often invokes dualism, as it is taken to include bodily processes, and nature 'out there'. Both senses of 'biology' bear critical scrutiny, as we shall see.

The underlying opposition between nature and culture emerged most strongly in the West during the Enlightenment; the study of 'man' became separated from the study of 'nature', and the boundaries of each were policed increasingly (Horigan 1990). While the emerging human sciences were defending their territory and carving intellectual space for themselves, the natural sciences came more and more not just to study 'nature' but also to adopt a particular philosophical position – reductionism – within that study. In the twentieth century, the success of molecular biology has driven the wedge further (Benton 1991). Allocating something to one side of this divide rests on that history; the workings of the human body, as well as the rest of nature, have generally remained on the biological side of the chasm.

In objecting to biological determinism of human behaviour, feminist critics have paid little heed to descriptions of hormones making changes to the biological body – to the lining of the uterus, for example. This is the realm of the biological, after all, the realm of biological facts: and it is rarely contested in feminist writing. Similarly, non-human animals (and the rest of nature) can be cast as pure biology, including their behaviour, while *our* behaviour somehow escapes into the ether. There is a little bit of wanting it both ways here. If we accept biological events as possible causes in relation to the biological body (as feminists indeed have in relation to health), then can we really expect to argue that biology is completely irrelevant to any discussion of gender?

Undoubtedly, biological arguments have all too often been made in ways that buttress gender divisions. Politically, then, feminists have tended to oppose biological arguments and to insist on some form of social constructionism of gender, or of other social categories (such as sexuality). But opposing biological determinism is one thing; throwing out discussion of biology altogether is quite another. In the zeal to reject biologism, the embodied subject and the biological body seemed to be forgotten. (Indeed, it is precisely that point that is often taken up by opponents of feminism, who usually insist on interpreting 'the' feminist position as being one of extreme environmentalism. In such a position, they argue, the reality of biology is completely ignored or ridiculed.)

As I write this, I have to confront the slipperiness of the word 'biology' and its multiple meanings as a branch of science, as the rest of nature, or as the processes by which our bodies operate. I am all too well aware that I will probably use it in different ways in this book (though I try to indicate its problematic nature). An important task for feminist theory – indeed, for all social theorising about the body – is to engage with the multiple meanings of 'biology', and of the 'biological' body, rather than leaving the category 'biology' largely uncontested.

Yet, apart from studies of the discourses of immunology (for example, Martin 1994), and Jackie Stacey's cultural study of cancer (1997), what happens inside the body does not seem to get out of the biomedical texts very often. It is left to the biologists. But by ignoring bodily insides, we run the risk of perpetuating the story of the biological body as fixed and presocial – even when that is apparently denied by arguments that we cannot understand our biological selves *except* through culture. Where does that leave (say) the action of nerves, the functioning of immune systems, or the development of embryos from fertilised eggs? That *level* of bodily working seems to remain forever outside culture, fixed into a 'biology' forever doomed to separation from society.

We must take 'the biological' more seriously; and we can do so (as I argue later) without lapsing into the dangerous waters of biological determinism. When we have time to get off the merry-go-round of necessarily responding to yet more determinist stories, feminist biologists have also sought to think about ways in which 'the biological' might be integrated into our theories. So, for example, rather than deny that hormones or genes (say) had any role at all, we have emphasised the complex ways in which they could themselves be influenced by, or act in concert with, our environment (see Bleier 1984; Fausto-Sterling 1985; Birke 1986; Hubbard 1990, for examples).

It is possible, even necessary, these authors have argued, to think about biology in ways that are less reductionist. Critics insist on viewing nature/biology as more complex and less fixed than many reductionist accounts allow, and they insist on *interactive* models of causality. One example comes from evolutionary theory. Critics concerned with ideas in evolution often deplore the predominance of notions of competition in nature, and of 'selfish genes' in modern evolutionary theory. Among other things, critics point out, such discourse is often both rooted in, and makes justificatory claims for, a particular form of late capitalist society (for example, Bleier 1984).

Such conceptions of nature assume an organism's environment to be dead and static, providing merely a set of circumstances to which to adapt. In her critique of modern evolutionary theory, Judith Masters (1995) notes how ecological and evolutionary theory speaks of 'niches' which have to be 'filled' by species – surely a passive notion. She draws instead on other ideas from studies of evolution, which represent environments as themselves composed of interacting organisms who change all the time, thus changing 'niches'. Organisms and their surroundings co-adapt, she insists, rather than one adapting *to* the other as conventional neo-Darwinism would have it.

Such themes – opposition to reductionism, and a concomitant concern to infer different accounts from available evidence (in less reductionistic frameworks) – run through most feminist work on biology. Such work insists on looking at nature, at biology, as much more complex and as less passive than they are often seen – both within popular scientific narratives and within the social sciences. A central point of this work – and one I want to emphasise strongly – is that what we draw from the study of biology need *not* always be determinist; 'biology' need not (and indeed does not) impose the limits on human potential that are so often feared within social theory (see also Rose 1997a).

These themes have run through most of my work in feminism and biology; they have informed both my feminist criticism of science and also my research in biology over the years. They also structure this book, as later chapters will make clear; my starting point for thinking about the biological body is my commitment *not* to lose 'biology' altogether in our collective feminist zeal to analyse bodies as socio-cultural products. On the contrary, I seek to examine ways to bridge the chasm between social theory and the understandings of bodily workings that have been thrown into the ragbag of 'biology'. The next chapter turns to the 'body question' in social theory, looking at the focus of such theory on the external surfaces of the body. The inside, as we shall see, becomes, by default, a kind of empty space.

2

Black Boxes and Tedious Universals: Feminism and the (Biological) Body

> The homeostatic animal is essentially like a self-regulating machine, and the characteristics of such a machine provide a useful basis for understanding the activities of the living organism. (Berrill 1970, p. 536)

The body-machine – a familiar image. These machines were everywhere in the textbooks of my student days. They were universal – 'the' animal body that stands for all kinds of bodies. They were also constant, maintained in balance by homeostasis. This image of the body is very different from the kind of body to be found in much recent social theory. There, what predominates is an image of the body as surface, always malleable and subject to multiple readings.

In this chapter, I want to examine some of these ideas about the body in social theory, particularly in feminism, in order to question suppositions about the biological body. The chapter begins with an overview; I examine how feminists have, over the last two decades, conceptualised bodies. 'The body' certainly enters feminist theorising now: but, I suggest, it remains one whose interior processes are rarely called into question (except in the somewhat different sense of interiority implied in psychoanalysis: see Marshall 1996).

From there, the chapter moves to more recent theorising and what I perceive as the limitations of much of this theory with regard to the 'biological body'. I then consider some of the ways in which gender has been read onto this body in scientific accounts, in the light of feminist critiques, and at how feminist work has tried to prioritise alternative ways of 'thinking the biology of the body'. This will set the scene, as it were, for a more detailed focus on bodily insides, and on feminist revisionings, in later chapters.

What is it to be embodied? The sex/gender distinction of the 1970s stressed gender, leaving embodied 'sex' largely untheorised. Indeed, Elizabeth Spelman has suggested that Western feminism has been infected with somatophobia, or fear of the body. This has been caught from centuries of Western somatic fears associated with gender. 'Women', she reminds us, 'have been portrayed as possessing bodies in a way men have not. It is as if women essentially, men only accidentally, have bodies' (Spelman 1988, p. 127). Feminism has tended to be somatophobic, she argued, because all too often the link between women and our bodies has been seen as curtailing our freedoms. Not only that, but similar biologism has been used repeatedly to curtail the freedoms of other 'others' – those who are black, or lesbian, or physically disabled for example. Spelman pointed out that dealing with bodily functions (especially those of others) has typically fallen to women and/or black people. 'Superior groups, we have been told from Plato on down, have better things to do with their lives', she points out (ibid., p. 127), like pretending they have no bodies.

It is not moreover, only the need to take care of bodies that defines women, but also the alleged determination of our behaviour *by* our bodies, and the values placed upon that behaviour. That is, some bodily organ or attribute – be that possession of a uterus, the vicissitudes of particular hormones, or inheritance of certain genes – has at some time been seen as sufficient explanation for women's subordination. History also has many examples of biological arguments made about those groups with power (see Schiebinger 1993). But in those cases, biology is said to produce superiority, domination, while for women it allegedly is the cause of our inferiority.

Yet, in emphasising social constructionism, in opposing it to bio-logical determinism, we have perpetuated the dualism; and have played down the importance of the biological body itself. Like many women, I have trouble thinking about theories of social construction that ignore my bodily pain and bleeding, or that ignore the ways that desire (however constructed) finds expression through my material body. My experiencing of both pain and desire is both materially and culturally constructed.

Bodies, then, are troublesome. Despite their (re)discovery by social theorists in the last two decades, they remain problematic. How do we begin to think about 'the body' as having an inside – moreover, as having an inside that does things, all by itself? What kind of languages do we have to talk about it? My starting point for this book is that, while the newly emergent social theory generates all kind of interesting ideas about how culture is written onto the body, we are much more limited when speaking about the body's insides. Indeed, we often have to fall back onto the language of biomedicine, to the body–machine, if we wish to describe inner processes at all.

We can clearly draw on the language of biomedicine, to describe the anatomical structures, to narrate the tales of how physiological systems work. Or we might resist these tales, at times referring to our experiences of pain perhaps; sometimes we might prefer to turn to other systems of explanation drawn from popular culture or from a range of alternative medical systems.[7]

Alternatively, we might speak of interiority by considering the 'lived experience' of the body. We might approach this using either psychoanalysis or phenomenonology (see Grosz 1994). Examples include Elaine Scarry's insightful study of the inarticulacy of experiences of pain (Scarry 1985), and Helen Marshall's notes on the phenomenology of pregnancy (Marshall 1996; and see Diprose 1994). Yet, useful though such approaches undoubtedly are, we somehow seem to keep them in a separate box from the language commonly used to describe bodily functions.

Emily Martin's 1987 study, *The Woman in the Body* described, for example, how women might use a 'scientific' account of bodily events in some contexts, while resorting to phenomenological accounts (based on experience) in others. What was striking about her study was the way in which women often *resisted* the biomedical accounts, or mixed them up with experiential stories[8]; implicitly, they contrasted the largely negative scientific accounts of women's bodies to more positive stories. Yet the stories of biomedicine are powerful, easily dominating our discourse. How else can we describe what we think is going on inside our bodies?

By contrast, the interior as biological body has been largely ignored in recent feminist theorising about the body; it has been relegated to the category of 'scientific'. Social theory has tended to focus on the exteriority of the body, on which culture becomes endlessly inscribed. Implicit in this is the related assumptions that 'the biological' lies outside social theory. Alongside the new developments in cultural understandings of the body as surface – malleable to intervention through piercings, tattooing, cosmetic surgery and so on – the body's interior lurks, unquestioned. It remains, suggests Margaret Lock, '"black-boxed"

– a tedious universal, and therefore consigned to the biological sciences' (1997, p. 268). That neglect of the biological also hides a wider politics; as Terry Eagleton (1993) wryly observed, 'there are mutilated bodies galore', in recent theory, 'but few malnourished ones, belonging as they do to bits of the globe beyond the purview of Yale' (p. 7). We can make cultural choices in the West to pierce our bodies or to undergo cosmetic surgery, to make our bodies different, precisely because we can take so much of the biological necessities of living, of maintaining bodily integrity, for granted.

Feminist theorists have analysed in depth the concept of difference, and how that is inscribed on the body: what constitutes and produces difference, for example? Difference here means not only gender, but also the multiple social differences of race, class, sexualities, degrees of impairment/disability, and so on. Some differences seem, indeed, to be inscribed upon the surface; others are less so. Yet the anatomical, internal, body in these formulations seems to disappear, except as a set of signifiers. Caddick (1992) argues that earlier feminist theorising that minimised sexual difference (insisting on androgyny), and recent feminist explorations of difference, have in common that they disembody. Bodies, in some theorising, seem to become little more than texts, she suggests.

Within the narratives of biomedicine, by contrast, celebration of difference is minimal or non-existent. Rather, the textbook descriptions of bodily interiors and machine-like physiological systems assume a 'typical' or 'normal' individual, whose systems are depicted. Physiologies that depart from these norms generally enter the realm of the pathological, not the worlds of 'difference'; these are the machines that break down. Whatever differences we might speak of between people – in their culture, or in the external appearances of their bodies – their interiors, we assume, are similar. The only differences that are consistently produced in the texts of modern biomedicine are those between 'normal' and 'pathological'.[9]

The assumptions that 'the biological' means fixity and normality, run through much social (and feminist) theory. But they are also products of the development of particular ideas in biomedicine, especially as these emerge from the labs and enter popular culture. Physiology is particularly prone to that kind of interpretation. Concepts of homeostasis, of the maintenance of constancy in the 'normal body', are one example; pathology and illness represent failures of that control. To a physiologist, the body dynamically creates the (apparent) constancy. But the dynamism seems to disappear in more popular accounts. This,

27

combined with the relegation of 'biology' as outside the realms of social theorising, serves to fix the internal body. And out of that fixity comes the determinism to which feminists are inevitably opposed.

THE DEVELOPMENT OF FEMINIST THINKING ABOUT BIOLOGICAL BODIES: AN OUTLINE

The biological body has long been a problem for feminism. Some nineteenth century feminists, such as Elizabeth Cady Stanton, argued a political position based on natural rights and similarity between women and men. But the influence of Darwinian ideas of human relationships to nature strengthened throughout that century, giving rise both to biological determinism used against women (the infamous argument that women's reproductive health would suffer, for instance, if they went into higher education) and simultaneously to feminist arguments for difference (rooted in notions of separate spheres: Rosenberg 1982). Antoinette Brown Blackwell, for example, writing in 1875 used ideas from evolutionary theory to argue against biological determinism. She did so by noting that the determinist arguments propounded so often by medics assumed inferiority for women. Blackwell, however, allowed that male and female bodies were reproductively different, but insisted that 'nature', through evolution, had given each specific strengths: thus, bodily difference, suggested Blackwell, does not imply that inferiority is biologically inevitable (Blackwell, in Rossi 1973).

In this section, I want briefly to sketch ideas about the biological body found in feminist writing over the last few decades. Some earlier feminists, such as Simone de Beauvoir and Shulamith Firestone, have often been interpreted by later writers as adopting forms of biologism. To an extent, I agree with those interpretations; yet, there is a danger that in invoking biologism, we throw the baby out with the bathwater, so to speak. For what some of these earlier feminists were trying to do was precisely what many of us seek to do now – to retain an understanding of woman's body as material along with our analyses of gender as sociocultural. So, in sketching out some of these lines of thought, my aim is to trace this earlier history of the biological body within feminism.

Simone de Beauvoir's influential work, *The Second Sex*, published originally in 1949, was rather ambivalent about woman's relationship to her body. On the one hand, she famously asserted that 'one is not born, but becomes a woman'; the category 'woman' thus seems to be a constructed one. She also emphasised that different women will experience their biological bodies differently (middle class women might experience the menopause differently from peasant women, for example), thus making clear the possibility that our experiences of the body are themselves socially contingent. In that sense, de Beauvoir's vision is one that matches that of many recent feminist writers. At best, the biological body is seen as something which can be experienced differently by different individuals.

Yet, on the other hand, her vision is problematic (see also Okeley 1986). For although she minimises the role of biology in many places, she also claims that: 'man, like woman, is flesh, therefore passive, the plaything of his hormones and of the species, the restless prey of his desires' (de Beauvoir 1969, p. 460). Transcendence above mere bodily functions is the mark of true humanity, she contends; but women are too likely to be stuck in immanence, bound by bodily dictates through reproduction. 'Biology' is thus passive and limiting in at least some of de Beauvoir's work, yet paradoxically active in the sense that women's reproductive organs seem to 'have a life of their own'. De Beauvoir's vision of the body is problematic partly because she relies on an additive model of the relationship between 'biology' and social experience. I have criticised additive narratives in detail elsewhere (Birke 1986, 1992); put briefly, these are accounts which imply that there is a biological bedrock onto which experience and environments write. Feminism cannot hope to change the bedrock, but we can concentrate on challenging the political system which structures experiences. In doing so, of course, the material body is ignored.

That turning away from the material body underlies a problem in feminist critiques of biologism, which is that *any* mention of the 'biological' body has sometimes been greeted with claims of biologism or essentialism. That has certainly been my own experience in trying to speak about biology and the body over many years; the 'body' I always wanted to discuss was not only a socially and culturally constructed one, but it was also material, it was flesh. It hurt, it bled. What also hurt was the antagonism I remember from other feminists in the 1970s because I dared to mention the great unmentionable: the *biological* body. The biological body remains an embarrassing theme.

Unlike de Beauvoir, Shulamith Firestone is usually portrayed in recent feminist writing as a biological determinist. Pregnancy, she averred, is barbaric; the answer to women's subordination thus lies in future technological control of reproduction. While she did indeed make those claims, I want partially to reclaim her work here. It was, of course, a product of its time; the prose is straight from the revolutionary Marxism of the 1960s. But it is also a text *ahead* of its time. In discussing, for example, the wide gap between empirical, positivist science and what she terms aesthetic culture, she suggests that a new cultural revolution might begin to breach that boundary. Rereading her 1969 book in the mid-1990s brought to my mind discussion in feminism of cyborgs and cyberfeminism, of breaching boundaries through new genetics, which seem to me to threaten the borders in much the way that Firestone implied.

Imagining many of the developments in genetics and assisted reproduction which have since become reality, Firestone moved on to her famous assertion about barbarity, noting also that 'childbirth *hurts*' (1979, p. 189; her emphasis). My reading of this, however, is not one of latent biologism, but one of saying that bodily experiences are tangible, yet themselves structured by society. Her subsequent advocacy of technological childbirth is made in the context of her utopian envisioning of a future in which women, too, have control over technology's uses; this includes domestic chores (which become centralised) as well as making babies.

Feminists have rightly been concerned about the progress of reproductive technologies, and women's loss of control over their own reproduction. But Firestone's vision was of women having (more) control: moreover, it was one in which the body and its limitations was itself subject to transformation – albeit through technology. In that sense, her view of the body as potentially *transformable* is important. All too often, the body in feminist theorising becomes transformable only on the surface. To suggest that the biological body is transformable is very different from saying that it is determining. In that sense, Firestone's work was not biologistic.

Moreover, recent theoretical works tend to ignore the contributions of feminists writing earlier. Elizabeth Grosz's influential *Volatile Bodies* (1994), for example, makes no reference at all to earlier feminist work on the body – for example, that of Adrienne Rich (1976). Rich's work is often interpreted as being biologically determinist, as she asks women to reconsider their relationships to their bodies, to female biology. Women are alienated from our bodies, Rich suggests (in similar vein to Spelman's 'somatophobia'): 'I know no woman ... for whom her body is not a fundamental problem' (Rich, ibid., p. 284).

Her answer is that we must learn to 'repossess' our bodies. 'In arguing that we have by no means yet explored or understood our biological grounding, the miracle and paradox of the female body and its spiritual and political meanings, I am really asking whether women cannot begin, at last, to *think through the body*, to connect what has been so cruelly disorganized – our great mental capacities, hardly used; our highly developed tactile sense; our genius for close observation; our complicated, pain-enduring, multi-pleasured physicality' (ibid., p. 284; emphasis in original).

I want to focus for a moment on the statements Rich made about the biological body. While I would dispute her supposition that we can understand our biological 'grounding' (precisely because it posits biology as presocial, *as* a grounding), her advocacy that we learn to 'think through the body' seems to me to come quite close to more recent work on the 'lived body' (see below). Her argument, moreover, that we should not throw out all reference to 'biology' in feminist theorising is one which feminist biologists, writing somewhat later, have also emphasised. It is precisely her desire *not* to lose the biological body from feminist theorising that I want to emphasise. Indeed, if Rich has been read so often as being biologically determinist, this may have much to do with how 'biology' is interpreted.

The 'biological' has, as I noted in the previous chapter, become a kind of ragbag into which a highly heterogeneous and eclectic mix of stuff can be thrown. All too often, it stands for all that which is not assigned to the categories of social and human.

BODY WORK

Social theory began to (re)discover the body in the 1980s. In particular, theorists began to address the significance of the body in consumer culture as having symbolic value; as such, it is indeed the surface, the outside of the body that carries that value, and by means of which we increasingly evaluate our sense of self (see Turner 1992; Shilling 1993, p. 3; Lock 1997). Yet only rarely does such theory move beyond the surface, remaining, as Rosi Braidotti has dubbed it, 'exteriority without depth' (Braidotti 1989, p. 154).

Surfaces, moreover, can be changed; they become so malleable, indeed, that they become an expression of the self – or of the mind. We can express ourselves (or not) by altering the body surface, temporarily or permanently. In at least some recent sociological theorising, people have become essentially minds, their subordinated bodies coming to consist of 'thoroughly permeable' flesh that can be redesigned at will (Shilling and Mellor 1996).

The focus on the body as surface has not gone uncontested, how-

31

ever. Reflecting on his own experience of the 'lived body' in illness, Arthur Frank (1996) noted how strange the concept of bodies as exterior inscriptions seems to someone suffering from cancer (see also Stacey 1997). 'To consider the body as pure surface sacrifices too much reality for the resulting theoretical convenience', Frank notes (ibid., p. 56). For him, illness 'becomes *experience*, which I understand as the perepetually [sic] shifting synthesis of this perpetually spiraling dialectic of flesh, inscription and intention' (ibid., p. 58; his emphasis).

Yet, beyond experience, the inside of the body seems empty. Writing about shifts in body-images in Western society, Ferguson (1997) describes a transition in ideas of the body from modernity to post-modernity. Here we find the elements of the absent internal body, subsumed into a kind of empty space. The modern body, Ferguson suggests, is a closed and impenetrable object, which emerged concep-tually as the body became identified as the possession of unique persons during the Enlightenment. This body is not connected socially, but is hermetically sealed: indeed, the outside becomes 'a thin, hard surface which served to separate an empty interior from an empty exterior' (ibid., p. 8). Emptiness inside mirrors our isolation.

Postmodernism has been particularly influential in recent body theory, lending an apparent fluidity through its rejection of the apparent certainties of modernity (including those of science).[10] 'Corporeal flows' (Grosz 1994) predominate in this theory, implying a dissolution of the boundary between inside and out. But this can also become, as Ferguson puts it, a conception of the body as 'an extended *surface* upon which, on the one hand, the externality and objectivity of nature, and, on the other hand, the unique inner quality of the soul, have jointly condensed' (ibid., p. 11). If the body is extended surface upon which culture writes itself, where then is the inner physiological body? It remains, as he later points out, 'uncannily empty' (p. 24) or a 'postmodern "body without organs"' (p. 28, paraphrasing Deleuze and Guattari 1987).

To think of corporeal flows and fluidity is certainly useful, and I will pick up that theme (and its limitations) in later chapters. Post-modern theory has, moreover, encouraged us to think in terms of the potential transformability of at least the body surface. This in turn implies some degree of control 'over' nature; indeed, there is no pregiven 'nature', rather, it is always culturally contingent and change-able. But these positions seem to be predicated upon the neglect of the interior body, indeed upon an assumption that it stays the same, looks after itself – or worse, that it might get out of control and have control over us, through HIV infection, or cancer (see Lyon 1997; Stacey 1997a).

Feminist analyses of the body have, in general, taken as a starting point the control exerted over gendered bodies through social processes – by legal and institutional structures, for example, or by physical violence, or through advertising in consumer culture. Thus, several feminist writers have addressed the power of cultural images that construct femininity as coterminous with having a thin body: fat is indeed a feminist issue (Orbach 1979), precisely because so many images sell the ideal of thinness. To achieve or maintain thinness, moreover, requires *disciplining* the body, through dieting or through eating disorders (Bordo 1993). There are, of course, other ways of disciplining the (female) body, many of which have been the subjects of feminist analyses – body piercing, bodybuilding and cosmetic surgery, for example as well as the disciplining that comes through classifications of 'other' bodies (through race and skin colour, for example).

The theme of disciplining the body draws extensively on the work of Foucault as the 'father' of new theoretical work on body politics (Foucault 1973, 1979). This has not been without feminist criticism: Susan Bordo, for example, reminds us that the new 'politics of the body' also had a mother – feminism and its history of engagement with cultural constructions of the body (Bordo 1993). Nevertheless, Foucault's work insists upon the body as preeminently a site of political control, increasingly subject to surveillance. Practices of medicine, for example, began to change during the eighteenth century, as medical power became more consolidated; what emerged was 'a generalised presence of doctors whose intersecting gazes form a network and exercise at every point in space, and at every moment in time, a constant, mobile, differentiated supervision' (Foucault 1973, p. 31). Monitoring of the nation's health is now so taken for granted that we barely notice.

In practice, our bodily interiors are disciplined at every turn. Cultural proscriptions and prescriptions control what we eat, for example (Falk 1994; Lupton 1996). When we become more aware of our bodily interiors as a result of illness, controls are exerted through encounters with doctors, through the structure of the hospital ward, and by the visualisation technologies whose output determines whether we are 'sick' or not.

From the late 1980s, feminist work on the body began to deal increasingly with concepts of embodiment, linking embodied experience to practices of power and agency. In particular, this shift enabled the analysis of embodiment and cultural practice as offering sites of *resistance* to cultural norms (see Davis 1997). If some cultural practices can be seen as normative (dieting to achieve thinness to fit a 'feminine ideal', for instance), others have been interpreted as subverting social norms (thus, the development of 'ripped' muscles by a female bodybuilder may be understood as subverting norms of feminine body shape).

Much recent feminist theorising on the body draws on social constructionism, focusing on the ways in which gender is produced through cultural practices. Yet this focus all too easily loses the lived experiences of being embodied. So, for example, we might read about the gendered body in body-building practices, or in anorexia – but it is much rarer to read in these theoretical accounts about what it is (or might be) like to experience the body with ripped muscles or concave belly. The body, rather, becomes a passive recipient of cultural practices, denied even the agency of experience.

As a result, social constructionism tends to perpetuate dualisms of nature/culture, body/mind in the very process of its disavowal of the material. In efforts to transcend these dichotomies, some theorists have advocated more phenomenological approaches emphasising the *lived body* (for example, Young 1990). This body is not 'merely' biological, but is both signifying and signified, historically contingent and social; it becomes instead, 'a body as social and discursive object, a body bound up in the order of desire, signification, and power' (Grosz 1994, p. 19; and see Butler 1993).

Theorising lived experience is made difficult, however, by the Western philosophical tradition in which the body of 'Man' is taken as central and the norm. Difference can thus only exist as 'other' to that central figure. As several feminist writers have urged, we need to start *from* a position that theorises and assumes from the start different embodiments if we are to theorise the body around sexual difference (Diprose 1994; Shildrick 1997). Only by doing so, these authors suggest, can we begin to bring women's bodies and their specificities into discussion of ethics (particularly biomedical ethics).

These recent shifts in feminist theorising are important. Not only do they begin to challenge the prevailing underlying dualisms of the cultural versus the biological, but also they begin to offer us possibilities of developing alternative 'body politics' through transgressive practices (see Butler 1993; Davis 1997). Yet gaps remain. Apart from the failure to pay much heed to the body's material interior and its processes, there is no consideration of bodily development – how we get to become the bodies we are through our lives. Both of these fall largely within the remit of 'biology': as such, it has fallen largely to feminist biologists to begin the task of retheorising (see, for example, the work of Ineke Klinge (1997) on osteoporosis and the menopause). This task starts from an analysis of gender in nature, and the ways in which gender – as a set of sociocultural assumptions – is read onto the human body as it is described in the narratives of science. Perhaps nowhere is that clearer than in the assumptions underwriting concepts of hormones – molecular bearers of sex.

GENDERING THE BODY: SEX AMONG THE MOLECULES

The main focus of feminist studies of biology has, inevitably, been the ways in which 'nature' has been read as gendered. Here, I outline some of these analyses, in order to map out feminist critiques of biology and particularly of the gendering of the body in scientific narratives. The 'tedious universal' of the physiological body is not, on closer scrutiny, universal at all. 'The' human body, as it appears in the texts of biomedicine, is on the contrary a differentiated body: for 'universal', read 'man' and 'white'.

The history of science contains many examples of how scientists (and others) have read both gender and race onto the natural world. That is, they have seen nature 'out there' as illustrating their own cultural assumptions about social divisions. In turn, the gendered natural world so described provides plenty of fodder for those who would argue that the same cultural assumptions are natural. To researchers looking at primate societies in the 1940s, it seemed appropriate to focus on male dominance and aggression, and to see this as central to primate social cohesiveness; that model was, in turn, inevitably appropriated to justify or explain male dominance. And as more women entered biological research, and as feminism began to ask questions about gender, so the focus of primatology shifted; now, females became the centre of primatologists' attention (Haraway 1989).

Reading social differences onto nature is an old habit, but it began to develop in earnest in the West in the eighteenth century. As European expansion resulted in the 'discovery' of more species of animals and plants, and other groups of human beings, so European science became increasingly concerned with the demarcation of difference. Difference thus became increasingly inscribed into bodies – but always as difference *from* the perceived norm of male/European/human.

Several feminist writers have noted the significance of the ancient (Aristotelian) idea that woman was but an imperfect man. Women's inferiority lay in their bodily coolness, men's superiority in their body heat. In the work of the Greek physician Galen, the testicles were the most important organs in the body because they could cook the blood; women's testicles (ovaries) also did so, but less efficiently. As Nancy Tuana puts it: 'Woman remains, so to speak, half-baked' (Tuana 1993, p. 22).

The notion of women's imperfection persisted. Yet it existed alongside a belief that male and female were to some extent interchangeable. Thomas Laqueur suggests, for example, that the ways in which the reproductive organs of male and female were typically drawn in anatomical texts prior to the end of the eighteenth century implied interchangeability; women's uteruses were portrayed as penises turned inside out (Laqueur 1990).

It is not, he argues, that the illustrators got the anatomy wrong: what made them see the uterus and vagina as penis-like was a whole world view that saw sex as social rather than as organic. 'Sex before the seventeenth century', Laqueur suggests, 'was still a sociological and not an ontological category' (ibid., p. 8; and see Jordanova 1991). Anatomical texts depicted women's anatomy as being intrinsically similar to men's, even if imperfect. The similarity, moreover, enabled possible transformation; anecdotes abounded of women who developed a penis after a bout of coughing, as their uteruses turned inside out.

The one-sex model described by Laqueur was certainly not egalitarian; while it was believed possible for a woman to change into a man, it was not considered possible for a man's penis suddenly to turn shy and run inwards. To become a woman in such ways was unthinkable. Nor, as Margrit Shildrick has put it, was there any sense in which men were seen as emptier because their organs were outside (1997, p. 42).

To what extent the anatomies of women and men *were* considered interchangeable during this period is a matter of conjecture. However, argues Laqueur, the notion of two quite different anatomies developed strongly in the eighteenth century; where organs had previously been seen as similar in the two sexes, now they acquired different names. Female testicles became ovaries, and the vagina was named.

Moreover, it was not enough merely to rename the testicles. As Londa Schiebinger's analyses of gender in the history of science show, physicians of the eighteenth century sought the 'essence of sex' in every part of the body; differences in bones, hair, blood vessels, sweat and brains of men and women were catalogued. She documents how 'the' skeleton of earlier anatomies turned into distinctively male and female skeletons during the eighteenth century. Representations of skeletons were deeply imbued with cultural ideals of gender; thus, female skeletons might be represented in coy posture, and with a heart on the sternum. Male skeletons could be depicted with a virile, if skeletal, horse (Schiebinger 1989).

What was begun in the eighteenth century became firmly consolidated in the nineteenth. Scientific accounts of bodily differences indicative of gender and race abounded (Schiebinger 1989, 1993). There was an explosion of classification of species followed closely by classification of different 'types' of human body throughout this period, as Foucault (1979) has noted. Gradually, the concept of the body as *foundation* of gender and sexuality emerged (see Wijngaard 1997).

A concern with the origin of differences was fuelled by the publication, in mid-century, of Darwin's theory of evolution. Speculation abounded about the evolutionary relationships of humans to other animals, and between different types of humans. Some commentators even wondered whether women were as highly evolved as men. Differences were diligently mapped, and new species slotted into new systems of classification (Schiebinger 1993).

In the twentieth century, that concern has become molecular, tracing the patterns of difference to the genes that a person inherits. Once genes were posited as the inheritable 'factors' that had troubled Darwin,[11] the spotlight turned inevitably towards genetics. Genetics and molecular biology have become, by the end of the twentieth century, the growth industries of biomedicine. And not surprisingly, in light of the growing passion for molecules, the search for the 'essence of sex' in every corner of the living world came to focus on molecular structures. Gross anatomy was not enough truly to differentiate male from female. In the early years of the twentieth century, the sex chromosomes were named, so distinguishing the sexes; and more recently, scientists have announced a 'gene for maleness' (called the *Sry* gene – producing testis-determining factor so pushing embryonic development towards bodily maleness). Note, however, that the 'essence of sex' being read onto nature is, again, not one of equality. Rather, it is one in which maleness is prioritised; being female is the default option (see Fausto-Sterling 1989).

Reproductive physiology has had a chequered history. It was always controversial, for many reasons – partly because of its association with sex, and partly because of earlier associations with clinical quackery (Clarke 1990). The charge of quackery stemmed from attempts, at the end of the nineteenth century, to promote vitality or longevity through organotherapy – the use of body organ extracts; Brown-Sequard, a French doctor, for example used animal organs on himself, including monkey's testes, in experiments to determine their vitality potential (see Borrell 1976). Testes, again, become the elixir of life.

Controversy aside, the idea that hormones might be involved in the physiology of reproduction led to intensified research, particularly from the 1920s (Oudshoorn 1994). The concept of a 'hormone', secreted internally from the endocrine organs, was mooted in 1905 (Hall 1976), and led to research into various endocrine organs – the pancreas, the thyroid, the adrenal glands. The emerging science of endocrinology, alongside the growing focus on evolution and genetics after 1900 (and with it an interest in genetics and population), fed into reproductive physiology.

A necessary first step, then, was to link reproductive processes – already thoroughly absorbed into concepts of two distinct sexes – to molecules. A second step was to label these molecules explicitly as *sex hormones* – a label still used today to describe some of the steroid hormones, such as oestrogens and androgens (and sometimes also applied to hormones produced by the pituitary gland, which stimulate ovaries and testes).[12]

Nelly Oudshoorn (1993) has documented how sex hormones were categorised during the 1920s as sexually specific: 'female sex hormones' were produced by female bodies. Partly, this was a question of labelling, of calling a hormone 'female'. But partly, too, it was a product of the use of particular kinds of biological assays: hormones were defined as male in terms of their capacity to make a cockerel's comb grow, for example. That is, they become male because they make males … well, more male.

Once these molecules had been defined as themselves having gender, the stage was set for even deeper probing into the very 'essence of sex'. In describing this search further, I want to emphasise three points. First, biologists often use gendered language in describing nature; hormones associated with reproduction are not the only instance. Second, the naming of hormones as gendered makes assumptions about the relationship between hormones and bodies. Third, following from this, those assumptions underwrite how biologists have developed ideas about the development of gender and sexuality in embryos.

GENDERED LANGUAGE

Gendered language appears throughout biomedicine. In some areas – reproductive physiology, say – we might perhaps expect it. In a way, it is perhaps unsurprising that scientists write about eggs and sperm in ways that are thoroughly sexist. The Biology and Gender Study Group (1989) and Emily Martin (1987) have both noted the language in which gametes are typically described in textbooks: sessile, coy eggs await the virile heroics of sperm.

But cultural assumptions of gender and race thread through all areas of biology, even where 'sex' is not directly relevant. Cells in immunology can be portrayed as feminine and white, for example (Martin 1994), as can various molecules – and even bacteria (Spanier 1995). This is not to say that scientists are deliberately, or even consciously, using sexist language; rather, that language is readily culturally available. Gender, notes Bonnie Spanier in her analysis of molecular biology texts, is sometimes read onto even the tiniest entities – molecules, or bacteria. Note that she does not only mean 'sex', in the sense of having two sexes; what she uncovers is the genderising of microscopic or molecular life.

Spanier remembers her bewilderment when she opened a major textbook in molecular biology, and found there not only 'male' and 'female' bacteria, but also that what determined the difference was something the 'male' possessed that the female did not – a not unfamiliar story. The bacterium in question was the oft-used *E. coli* (which is to bacteriology what the white rat is to psychology). Scientists had for decades been observing the transfer of genetic material (a tiny circle called a plasmid) between two cells of *E. coli*. Cells with a plasmid were labelled male; those without (of course!) were female. Not only do the bacteria thus acquire gender, but the plasmid becomes a male signifier, Spanier notes. And not only that, but the male signifier is then labelled as 'an essential tool [sic] of recombinant DNA technology' (Spanier 1995, p. 58, citing a textbook of molecular biology).

Gendered molecules thus abound in nature, or so it would seem. Macho molecules and butch bacteria appear to lurk in the test-tubes.

SEX HORMONES AND BODILY DIFFERENCE

The sex hormones seem obviously gendered, in that they are associated with secondary sex characteristics – by which we might read a person's gender. The 'male' hormones, such as testosterone, stimulate beard growth, a sign that we usually read as male, while 'female' hormones such as oestrogens stimulate breast growth.

Sex hormones (belonging to the category called steroids) may be secreted by ovaries or testes, as well as the adrenal glands sitting above the kidneys, but – like all hormones – they move around the body in the blood. And they can cross the partial barrier between the brain and the rest of the body. If these molecules were intrinsically gendered, the reasoning went, then they must have a role in the differentiation of gender. And if they can get into the brain, then they must be able to affect behaviour. This is the basis of biologically determinist claims; hormonal differences in the body translate into differences in the brain and its function. From there, it is but a short step to inferring that, for example, men do not do the ironing or that women are better at housework, because of the way their brains are wired up.

The labelling of hormones as gendered belies their complexity. Indeed, the steroid hormones have far more in common with the transgressions of postmodernism than they have with the simple dualisms imposed by scientists. All of us produce all kinds of hormones, albeit in different quantities; steroids come from other glands besides ovaries and testes. And, to make things rather tricky for simple assumptions about the virility of molecules, androgens ('male' hormones) are often converted to oestrogens inside the cells, while progestins ('female' hormones) can 'masculinise' infant rats.[13]

So why the dichotomous labelling? Starting from the (problematic) observation that there are two different kinds of bodies, serving different roles in reproduction, the assumption seems to be that there will be some underlying bodily differences that map onto the external ones. If there is any mapping, it is a quantitative one; males tend to produce more of the so-called male hormones (the androgens) and females produce more 'female' hormones (oestrogens and progestins). This generalisation is, however, just that: it is a statement about averages.

Statements about 'sex differences' are ubiquitous, implying a qualitative difference. Yet almost none of these differences are absolute; indeed, the overlap is considerable between female and male bodies on most physiological measurements (such as heart size, the capacity of the blood to carry oxygen bound to haemoglobin, or blood volume: see Birke 1992). 'Difference' here is overdetermined; that is to say, that it is exaggerated out of proportion to whatever differences can be demonstrated scientifically. These are also generalisations across different ages; differences are much less in young prepubertal children and in older people, particularly past the age of women's menopause. But by labelling hormones as intrinsically gendered, endocrinologists have asserted a qualitative difference, hormones as having essence. Probing the essence of sex, indeed.

BECOMING GENDERED

Where age does enter the story about gender and difference, it is before birth. Not surprisingly, the sex hormones are part of the scientific story about how we become one sex or the other. Early exposure to hormones (long before we are born) ensures that the brain and internal reproductive hormones are pushed in one direction or the other; female or male, gay or straight. Brains, the story goes, are permanently organised by gender and sexuality by means of these hormonal washes.

One example of such thinking was Simon LeVay's much-publicised claim to have identified 'gay brains' (LeVay 1991). More precisely, he claimed that he had found differences in a small area of the brain (in the hypothalamus, near the base of the brain) between straight and gay men. The study was, however, widely criticised for both its methodological and logical flaws. Like many studies of alleged sex differences in brain structure, LeVay's study exaggerated what he saw as a slight difference, on average, into something much more significant.

Central to LeVay's claims is the belief that the brain has been permanently organised by sex hormones prior to birth. In the case of gay men, the bit of the brain involved was smaller than its equivalent in straight men; LeVay put it more graphically in a later book, when he suggested that gay men simply do not have the brains for straight sex (LeVay 1993).

LeVay's claims depend upon the organisation hypothesis, which contends that brain tissue is permanently altered by early exposure to hormones. In humans, that exposure occurs prenatally. The idea was originally proposed in 1959, based on studies of hormonal effects in animals. Although undergoing various changes, the hypothesis is built upon assumptions of dualism, that brains are pushed by hormones into one or other pattern (Wijngaard 1997).[14]

Now in drawing attention to this, and in critiquing it, I do not wish to imply that it is simply wrong. Hormones do get into brains, and do seem to have effects. Among these is the part played by androgens in permanently altering the potential of one part of the brain itself to affect hormonal patterns. So, the brains of females tend to be cyclic over a period of days (in women, that corresponds to the menstrual cycle), while in males such cycles are abolished.[15] The problem for feminist critics lies in how these actions are interpreted.

Gendered dichotomy is etched deep into narratives of our biology, particularly in the organisation hypothesis. Femininity and masculinity thus become products not of complex social negotiations and cultural learning, but of inevitable biological processes happening while we are

41

still in the womb. These in turn are the products of equally dichotomous molecules, already defined as gendered. Cultural conventions of gender have been read onto molecules which in turn have become the definers of the cultural conventions of gender.

Yet there is considerable overlap, even among non-humans. Among the rats so beloved of researchers in this area, for example, females engage in so-called 'male' behaviour and vice versa. Furthermore, there is plenty of evidence that it is not all down to the internal processes of hormones; for rats, too, socialisation plays a significant role in the development of gender (Moore 1982; Birke 1989). Yet the dualism, and its embeddedness in biological determinism, persists.

What underlies the biological claims is an assumption that the developing foetus is alone in the world (a representation that feminists have noted in other contexts, such as photography of foetuses *in utero*, which omit the context – the mother: see Petchesky 1987). But it is not. Its internal biology, including its hormones, is constantly interacting with the world around it, including the physiology of the mother (and her social and cultural world). Her hormones, and the proteins in her blood that can 'mop up' certain hormones, are all part of the complex picture. That is also true of rats, whose prenatal life is additionally dependent upon the responses of siblings in the uterus. The outcome, biologically speaking, is a complex product of all these influences.

The hormonal body, in the scientific literature, is a body deeply inscribed with difference. It inscribes differences of gender – becoming male with testosterone, for example. And it inscribes differences of sexuality, for what underwrites much of the biological research into the origins of homosexuality (there is, of course, no corresponding search for heterosexual origins) is the assumption that a gay man is less of a man. For lesbians, it is a case of being more manly. These old stereotypes are taken for granted in research. LeVay's story is no exception.

We can only speculate, notes Marianne van den Wijngaard (1997) in her analysis of the organisation theory, what research might have looked like if cultural conventions were not so dualistic or so embedded in assumptions about hormones. As she notes in her discussion, feminists have – rightly – been critical of these hormonally determinist theories. Part of that criticism has entailed an insistence that there are other ways of thinking the biological body, ways that focus not on determinism but on transformation.

BODY TRANSFORMATIONS

One reason, as I have noted, why 'the biological' has been absent from a great deal of social theory is precisely its relegation to the status of presocial, as given. Some feminist work has dealt with aspects of the biological body and its cultural representations – analyses of the immune system, for example, in the work of Donna Haraway (1991b) or Emily Martin (1994) – or have dealt with high-profile areas of biology such as the new genetics. Yet there seem always to be areas of 'biology' omitted from such cultural analysis – usually, the body's interior as a whole, or perhaps its cellular structure. Interestingly, it is also the workings of the interior that largely escape the attentions of mainstream philosophers of biology, who tend to focus on areas such as evolution or genetics. Physiology, by contrast, seems doomed to narratives of mechanism and reductionism, unsullied by philosophical critique or the challenges of postmodern cultural criticism.

Most of the time, the workings of our body – our physiology – seems constant; it is part of our bodily 'nature'. I may be tired (lack of sleep), or feeling a little dizzy (which I have learned to call low blood sugar) – but on the whole, my body maintains a reasonable balance. In biomedical texts, we can learn that the body's functions – physiology – is categorised into systems: the nervous system, endocrine system, immune system and so on. A central principle of how these systems work, we learn, is homeostasis, the body's ability to maintain a constant state. So, for example, body temperature is normally around 37 degrees Celsius, and levels of sugar in the blood usually (except in some disease states) remain within certain limits.

Those ideas of separate systems maintaining constancy become part of a wider cultural language, assumed and implicit even within accounts of the 'socially contructed body'. Constancy is normal; perturbations represent disease. It is a theme of this book that ideas of constancy and normality, and implicit fixity, of the biological body are reinforced by its exclusion from recent social/feminist theory *of* the body. The exclusion serves to fix them. Yet this presents problems, for the fixed body is, by definition, one that changes little. And that brings us to a point made by many feminist writers – that this image of the body is a *gendered* one. The unchanging body is a masculine one; that is, the masculine body is assumed to be relatively constant (itself a dubious assumption), against which female bodies, with their ebbs and flows of bleeding, become problematic (Iris Young has made the point that it is probably only young men who actually believe that their bodies are relatively unchanging; women, certainly, are not likely to experience their bodies in such ways: see Young 1984). Pregnancy, too, is a process of change and transformation – in this case, of both woman and foetus. Women, on the whole, do not necessarily experience their bodies as unchanging.

Yet it matters that we think about the biological body as changing and changeable, as *transformable* rather than as a 'tedious universal' machine. Accepting concepts of the biological body as relatively unchanging and universally representative continues to fix it into determinism; by insisting on thinking about 'the biological' in terms of transformation and change, rather than fixity and stasis, we might be able to develop a conceptualisation of the biological that is not rooted in determinism. That is a theme I pick up later in the book, but which I want to emphasise here, before embarking on more detailed explorations of physiology and the body in the next few chapters.

3

Short Circuits: Reading the Inner Body

The Female Body is made of transparent plastic and lights up when you plug it in. You press a button to illuminate the different systems ... The reproductive system is optional and can be removed ...

Each Female Body contains a female brain. Handy. Makes things work. Stick pins in it and you get amazing results. Old popular songs. Short circuits. Bad dreams. (Atwood 1994, pp. 90–2)

Rita is lying in bed one night when she realises she's lost her uterus. She does not remember its falling out of her, like change out of an overstuffed wallet. Her ambivalent feelings about having children, she thinks, may have caused the womb to shrink away and fall out of her like a kind of discharge, an escape from lack of use. (Foos 1996, p. 1)

What is inside the body? And how do we know what is there? In this chapter and the next, I want to turn to these questions. But in doing so, I have to draw, at least partly, on the descriptions of science. Our understandings, however partial, are informed by the discourses of biomedicine in the late twentieth century. Memories may be rusty, but most of us learned some human biology at school, or we have watched TV programmes featuring forays into the body, or perhaps read popular health magazines or pamphlets from the doctor's surgery. It is this that tells us what is inside the human body, whatever our experiencing of our interiority may be; the scientific narrative has authority, it is a powerful voice. Indeed, what other language might we have to understand what we have inside?

It is physiology and anatomy, as ways of structuring our under-standings of bodily insides, that are my focus here. Partly, that is because of my own background as a biologist; a large part of my under-graduate training centred on physiology, especially neurophysiology. I have discussed elsewhere some of the issues that training raised (notably about the use of animals: Birke 1995). Here, I want to look critically at some of the concepts of physiology in the light of feminist 'body work'. Arcane some of these concepts may seem, but they are part of a wider discourse of biomedicine usually acquired through popularised articles: and where else do we get information about our insides except through the predigested and simplified stories told in such places about 'your body'?

One of the ways I came to learn some anatomy was through a transparent plastic 'Visible Woman', much like the one invoked by Margaret Atwood in her story. I thus learned that bodies have a number of separable organs, which I had to paint in different colours to make them distinguishable. That lesson was then reinforced through my scientific training – though now the organs were made of flesh and I was required to remove them from dead animals. I also learned that such organs are part of 'systems', which serve particular functions in the body. These ways of thinking are now familiar – so much so that we cannot easily understand how bodies were conceptualised in the past. Yet one reason why Laurie Foos's book is amusing is that it mocks our scientific certainties that, unlike our predecessors, we do *not* believe that wombs can go wandering about. The notion that one could simply drop out of us, escaping because not used, seems absurd.

My purpose here is to look at how scientific ideas about bodily insides have developed historically – from the practice of dissection to the visualisation of inner structures by means of technology. From there, I move on to consider how our insides are represented in scientific diagrams. In drawing on historical accounts, I do not want necessarily to imply direct lineages, but merely to map out a terrain. Many histories of science and medicine, moreover, rely on a story of science as inevitable progress: the truth, like the good guys in the movies, always wins out in these tales. My purpose, rather, is not to take for granted the 'reality' of scientific stories or their histories, but to sketch the ways in which the scientific concepts are socially and culturally contingent.

I begin, in this chapter and the next, with structure, how we might learn about what 'bits' we have inside our bodies. From there, I want to move in later chapters to function – what are the predominant themes of scientific stories about *how* bodies work? I do not intend to review exhaustively scientific concepts of structure (anatomy) or function (physiology); rather, I explore some particular themes, looking at the social and cultural context of their conceptual development. One specific focus is the representation of the body's inside through scientific diagrams.[16] In looking at cultural contexts and how the science of our bodies is represented visually in diagrams, I seek to ask questions about the impact of these ideas on how we perceive or think about our bodily interiors. And, of course, I examine the ways in which these narratives are gendered, and the implications this has for feminism.

One of the themes I want to draw out from histories of anatomy is that of space. If the body's insides seem to disappear into empty space in so much social theory, can we turn to scientific accounts for a more solid story? Organs undoubtedly have solidity, even rigidity, in the dissected corpse, so surely there is the basis for filling up the empty space? Perhaps there is, but it is always counteracted by images in diagrams that convey, for most lay people looking at popular descriptions of anatomy, a great deal of emptiness. Space, it might be said, is what structures (some of) our understanding of the inner body.

Later, I will examine the abstraction of anatomical diagrams in 'popular' books and textbooks, which create an illusion of space within the body. Organs, represented by simple lines, seem to float free within the larger space. That this illusion of space does indeed inform how people might think about bodies was brought home to me by a conversation regarding the design of an anatomical model for an exhibition in London, based on medicine. The artists working on the design had expressed considerable surprise that the organs of the body, shown in the lifesize model, fit together so tightly – so tightly, indeed, that it took some effort to put them all back together once dismantled. The artists, apparently, had imagined the organs of the body surrounded by spaces – just as they are shown in elementary textbooks.[17]

FINDING OUT: VISUALISING INTERIORS

Beyond gazing at the cervix – a sort of half-way house between a woman's inside and outside – there is little that most of us can do to see 'inside'. Self-examination was, is, a radical act in enabling women to see what was hitherto hidden; it thus challenged medical power. But to see the inside proper we must rely on doctors. To be precise, we must accept their interpretations of images generated by machines – ultrasound images of the foetus, for instance, or computerised tomography (CT) scans of parts of our bodies. Most of these images are generated when we are ill, with the exception of foetal imaging in pregnancy; they are part of diagnosis. All require interpretation, for they bear little apparent relationship to what we may know about the location of our bodily organs. How, for example, do the complex lines of the electroencephalograph (EEG) relate to what our brains actually do? It is often an act of faith to 'see' what we are told is there – such as seeing 'your baby' on the ultrasound scan.

So how do most of us find out about bodily insides – if at all? For the majority of people, the body's insides are rather mysterious. There are various organs, located in approximate positions, which few could pinpoint with much accuracy. We may – or may not – be aware of the symphony of noises and gurgles emanating from the inside, and how they relate to sensation or function. Few of us, however, bother to listen to those sounds, nor to try to picture what goes on when (say) our hearts beat faster, or our guts move food along in the processes of digestion.

We live, moreover, in an age when few of us ever see raw internal organs. We no longer (thankfully) witness the spilling of blood and guts on the battlefield, nor the decaying body hanging from a gibbet, while the dismembering of animals takes place away from our sight and hearing, in special places called slaughterhouses. Even the products are carefully wrapped in plastic and labelled in ways that minimise their origins as bodily parts. Instead, to learn about bodily interiors requires learning the language and practices of medicine: a few learn those directly, as part of medical training (including dissecting human cadavers), while most people learn it predigested, as it were, out of textbooks and models.

Blood and guts are obviously part of how we might imagine our innards. To imagine this is to call up images from the butcher's shop (the piles of offal) or from half-remembered programmes on television. Blood, in the form of blood-coloured dyes, is a very familiar image on the screen, the result of dramatised shootings or stabbings. Guts spilling all over the floor is less likely. To recall that kind of imagery, we might have to invoke scenes from programmes claiming to take us on voyages through the human body. Even then, trying to remember such representations as I write, I can recall mainly film of beating hearts – programmes on the heroics of heart surgery, for example.

Those televised views of bodily insides produce specific kinds of images: glistening flesh, exposed to the sharp light of the studio. Exposed flesh gleams and seems to slip through the surgeon's hands, like an elusive fish. Perhaps because the heart is so frequently the organ exposed to such well-lit scrutiny, camera-exposed flesh seems to pulse. It is, moreover, brought to the outside through the medium of the camera, and exposed to light. Through that visualisation, the imagined inner spaces of the body become continuous with space outside the body, and the organs are no longer hidden away in our innermost selves.

Architectural spaces have also provided metaphors for thinking about the body and its interior; they provide material for illustrations in a number of popular textbooks, for example. Pictures showing the body as a house, or as a factory, are still commonplace. These metaphors have a long pedigree. Medieval texts, for example, described the body's insides in terms of spaces and architectural structures (Pouchelle 1990).[18] Such metaphors have persisted – as that example of the exhibition torso illustrates. They were invoked, for example, by surgeon Richard Selzer writing about the experience of doing surgery: 'For the first time we can see into the cavity of the abdomen. Such a primitive place. One expects to find drawings of buffalo on the walls' (Selzer, quoted in Scarry 1995, p. 88). Inner space is a predominant motif in conceptualising anatomy (see also Foucault 1973).

Other means of looking inside the body also suggest space. Think, for example, of those television documentaries that take you 'voyaging' into the human body. Following the fibreoptic camera, you can voyage along a Fallopian tube, or a vas deferens, or swim through the arterial system. Here, you become a miniaturised viewer; but what you see, what you enter, is a kind of oceanic space, precisely because it is the body's channels along which you travel – pulsating, glistening, an inner space that is finally opening up its secrets to the probing eye (or I). In life, of course, these channels are filled with fluid – miniature canals. But the moving miniature camera seems to disperse the fluid, creating impressions of empty space.

Space, then, is a powerful motif in the way we might conceptualise the structure of the body's insides. To speak about its function, how it works, we might invoke other metaphors, such as industrial metaphors to describe the organisation of that space. Although I doubt that anybody actually imagines their insides to look like a factory, or even to work like one, the factory metaphor has often appeared in popular accounts of 'how the body works' (see Martin 1987). Other metaphors, derived from engineering and from militarism, also circulate and inform how we think about the functions of our internal organs and systems. Militaristic descriptions, for example, peppered the stories told about the immune system by scientists and lay people alike in Emily Martin's study (Martin 1994).

49

Blood/guts/space to describe inner structures, and metaphors of the militaristic body to explain how it works; we can call on any of these to think about what goes on inside. They are not the only frameworks; but they are culturally significant ones, derived from the literature of biomedicine, helping to organise how we conceptualise our biological bodies. It is hard to imagine how else we might think about them. In this section of the book, then, I want to examine further such ideas of space/structure and their background, beginning with a brief look at the history of dissection.

THE MEDICAL GAZE: THE SPACE OF THE DISSECTED CORPSE

One way of learning about inner anatomy is to study diagrams in books, perhaps combined with manipulating models of the kind I had as a child. Another is to learn through dissection of corpses. Some of us did that through the corpses of animals (killed for the purpose of our learning: see Birke 1994). The few who undergo medical training will do so through cutting into the corpses of dead people.

Dissection of cadavers is the obvious way to discover what human bodies contain. But its practice developed significantly only since the late Medieval period, gradually replacing earlier reliance on learning anatomy by studying the written works of great men (sic). Dissection, as an investigative practice, was partly a constituent of the scholarly revival of the Renaissance, although it also had a shady, disreputable, side. Probing the body's insides meant entering the abject, exposing the horrors of blood and guts; corpses, moreover, usually came from the gallows.

'The body' of this period was also a microcosm of the larger world (the body politic is hardly a new idea). Dissection in the thirteenth century had been permitted as punishment (following the gallows) or for religious reasons, but not for the pursuit of knowledge. Its practice was fraught: it 'brought symbolic dangers not only on the deceased and on the natural order, but also on the whole of society. If the members and organs of the microcosm were dispersed, disturbing the order of the world, would not the "body politic" fall apart as well?' (Pouchelle 1990, p. 83). Those fears – of the dangers of the body, of dissection as punishment, and of the dangers of disturbing the social fabric – persisted. (Indeed, they surfaced again in the early nineteenth century riots at the time of the Anatomy Acts in Britain; Richardson 1989).

The discoveries yielded by dissection were thus embedded in social and cultural change. This is illustrated by the Renaissance anatomist best known through modern medical histories – Andreas Vesalius of Brussels. Vesalius published his *De humani corporis fabrica* (On the fabric of the human body) in 1543, based on extensive studies of the anatomy of human corpses. His work is illustrated with detailed drawings of the internal organs, shown as they are located within the body, and located in a finely drawn setting (including the gallows).

Whatever his skills as a cutter of flesh, Vesalius was also a representative of a new style of scholarship – an explicitly masculine one. Historian of science Londa Schiebinger, in her excellent study of gender and the origins of modern science, *The Mind Has No Sex?* (1989), describes the 'battle over scholarly style' as a confrontation between feminine and masculine styles of presentation. Not only were male scientists likely to illustrate their work with pictures of themselves (unlike the few women practicing science in the period), but they were part of moves towards creating a self-proclaimed *masculine* philosophy in the development of science. Philosophy, explains Schiebinger, referring to the origins of the Royal Society of London, 'was to be distinctively English (not French), empirical (not speculative), and practical (not rhetorical). Consistent with the discourse of the day, each of these favored qualities was considered masculine' (ibid., p. 138).

Vesalius was one of this new masculine breed emerging in Europe in the sixteenth to seventeenth centuries. He broke with previous traditions, in which a professor read from a textbook while a demonstrator cut into a body on the mortuary slab. These changes marked him as radical to his contemporaries; his approach can be read as gendered, as the following quote illustrates:

> His basic reform there [Padua] was to do away with 'demonstrators' and 'ostensors', in the old sense, and to put his own hand to the business of dissection ... His demonstrations were on the human body, and he used living models on whom he marked the outlines of the joints and of other parts ... The work was carried on with great energy and drive. Every line and every figure of his great book, the product of but five years' activity, is instinct with his virile power. (Singer 1957, p. 114)

This passage, from a history of science text, is replete with notions of science as progress, with conveying the image of Vesalius as an innovator, a truly modern figure. Progress, moreover, is symbolised by Vesalius' drawings, 'instinct with virile power', a virility inspired by cutting into the passive flesh of the corpse. Entering the body became a masculine pursuit in more ways than one.

Whatever claims Vesalius might have had to being a great anatomist, however, his 'virile power' prevented him from accepting details of women's anatomy. He explicitly rejected, for example, the 'redis-covery' of the clitoris by his contemporaries, arguing that the clitoris was a 'sport of nature' found only rarely: 'you can hardly ascribe this new and useless part, as if it were an organ, to healthy women', he wrote (quoted in Park 1997, p. 177). Unable to identify a function, Vesalius felt obliged to classify this 'sport of nature' as a pathology.

Entering the body and describing its organs was very much a classifi-catory pursuit. Increasingly, the study of nature meant the classification of nature. By the eighteenth century, detailed systems of taxonomy were devised (Schiebinger 1993) – not only to categorise other kinds of organisms discovered during European expansion (including other groups of humans), but also different kinds of disease and symptom (Foucault 1973). All of these categorisations contributed to the development of modern concepts of the body. Dissection contributed to, and was part of, this move, enabling a shift towards seeing the body in parts, as assemblages of organs.

Vesalius had earlier referred to the *fabric* of the human body in his work; the body's parts were to be understood in the whole context, in accord with contemporary lay perceptions of ebbs and flows. This may be one reason why his illustrations are not diagrammatic, in the way we have come to expect, but 'posed as in the living body, and given a background such as that to which they were accustomed during life.' (Singer 1957, p. 116).

Yet dissection involved cutting *into* the fabric of the body, and separating out its parts. In that sense, it contributed inevitably to the notion that each organ has a specific, and separable, function (an idea that is clear in William Harvey's seventeenth century accounts of the circulation of the blood: Singer, ibid., pp. 174–5). That conceptuali-sation is with us still, and is central to our thinking of the body as a series of systems.

Indeed, we are so used to this perception of the body as a collection of bits, that it is difficult to imagine the inside of the body in any other way. Dissection was not only a question of discovering where different bits fit. It was also necessary to learn to read the internal body *as* a set of parts – and to interpret *how* the bits fit together.

In earlier centuries (up until the end of the eighteenth century), examination of a sick patient had relied largely on visual observation of external bodily signs and accepting what the patient said about her/his condition. Doctors' interpretations depended heavily on the humoural doctrine, according to which there were four 'humours' which must be in balance for health. Symptoms of disease thus depended on the individual's humoural constitution, and treatment was aimed at correcting those specific symptoms to regain equilibrium.

Accordingly, what mattered was bodily flows; both lay people and medical practitioners concurred in their understandings that flows were necessary for health, to rebalance the humours. Indeed, not allowing a wound to suppurate, and hence to allow the evil within to escape, was dangerous; to regain health required a 'symbolic exteriorisation of the trouble', to rebalance internal fluidities (Pouchelle 1990, p. 59). Barbara Duden, in her fascinating account of women patients in eighteenth century Germany (Duden 1991) explores the complex ways in which women spoke about their bodies. Her analysis focuses on the written records of a doctor, Johannes Storch, in the town of Eisenach around 1730. It is extremely difficult, she notes, to read and understand what the women of Eisenach (as written in Storch's records) thought about their bodies. We have to let go of our modern perception of the body as insulated from its outside, as 'belonging' to the self, if these women's stories are to make any sense to our readings.

The body in earlier cosmologies was not separated from its context; it flowed into and out of the social world. Duden draws on accounts of the plague in seventeenth century Florence, arguing that 'plague was perceived as a disease not only of the biological body, but also of nature and the social body' (ibid., p. 11). A crucial part of this view was a notion of a constant exchange between inside and outside of the body, such that curing disease meant trying to calm that exchange or rectify its imbalances.

The women of Eisenach similarly believed in the need for the body to flow; blood or pus had to be drawn out if stagnation and illness were to be avoided. Bloodletting was necessary, as were various means of luring the blood from one part to another (such as heat or herbal prescriptions). Storch records many instances of women suffering from such phenomena as 'suffocation of the womb' caused by an excess of blood surging towards uterus or head (ibid., p. 146); there, it caused stagnation and cramping, and must be lured away.

These women clearly understood their bodies in ways very different from our perceptions of self and body at the turn of the millennium. They were, moreover, living at a time of transition. No longer flowing quite so readily between biological body and social nexus, the flows within could more easily stagnate. Thus, nature 'withdrew into the body of women ... [i]t was only in the women's inner body that the doctor could listen to nature ... [Storch's] nature no longer knows anything about an order in the microcosm that corresponds to an order in the macrocosm' (ibid., p. 177).

Earlier, Medieval, concepts of bodies as flowing with exuberance were, by this time, themselves in flux, Duden argues. How different were the women of Eisenach from earlier depictions; they were 'wretched, stagnating, miserly (the richer ones), and cramped up – this is how they experience themselves, helpless in their suffering bodies, which were supposed to flow and did not. The agony of experiencing an objectified body had begun' (ibid., p. 178).

Although educated practitioners such as Storch must have known of Vesalius' work, the practice of anatomical dissection seemed to have little impact on clinical practice. '[T]he dead body did not yet cast its shadow on the living body', argues Duden (ibid., p. 106). Even while the dead body might be depicted as consisting of geometrical spaces, the organs, the living body was much more permeable and open. Thus, pain or a rash might 'occupy' a person, and make itself known by erupting in some part of the body; the concept of diseased organs was not part of this cosmology.

Eyewitness stories were more believable to Johannes Storch, than evidence from anatomy, even though the stories that people told were often anatomically impossible. Such accounts from living bodies were more convincing than dead ones. But, Duden argues, 'anatomical discoveries, like all of medicine up to the end of the eighteenth century, were still compatible with the miraculous that was slow to disappear' (ibid., p. 70).

Indeed, not all of those associations *have* disappeared. The phrase 'it made my blood boil' is hardly a reflection of modern physiological accuracy (we would not be here to tell the tale if it had). Rather, it recalls much earlier narratives of fluxes and flows, and their linkage to emotions such as anger, as well as metaphors of parts of the body in terms of ovens and cookpots (see Pouchelle 1990). Moreover, the images of space within the body, that I noted above, permit an understanding of the body in terms of fluxes and flows. It is, then, little wonder that – in contrast to the 'scientific' accounts of the physiology of menstruation – some women still speak of their concerns about blood 'having to go somewhere' if they experience lighter menstrual flow.[19]

That earlier worldview, of both nature and the internal body as fluxes and flows, began to change. Accompanying these shifts in

perception was a growing tendency to extend the external classifications of nature into the body. Where Vesalius had seen gender
difference only in terms of the reproductive tract (Schiebinger 1989),
by the end of the eighteenth century, anatomists were seeking difference throughout the tissues of the body. This, argues Schiebinger,
constituted a revolution in views of sexuality: 'By the 1790s, European
anatomists presented the male and female body as each having a
distinct telos–physical and intellectual strength for the man, motherhood for the woman' (ibid., p. 191). In accord with these diverging
concepts, the internal anatomies of women and men became
categorised increasingly in terms of difference. Muscles, sinews, bones
– all became gendered.

To classify requires seeing the units as separate – this is an X, that a
Y – even if the units belong together to some larger category, such as a
genus or a body. That process of separating nature out necessarily
helped to break down perceptions of the fluxes and flows of nature.
There was a conceptual splitting apart in taxonomy that mirrored the
literal splitting apart of the dissection table or (later) the physiology
experiments.

It was not, of course, only gender that was read into the inner
recesses of the body. Race, class, and sexuality were, too. The most
familiar examples refer to the outside of the body, to alleged markers
of difference or inferiority. But those constructions have also been
extended to the internal functions of the body, leading to claims, for
example, that blood from those who are black differs from that of
white people (with concomitant implications of contagion). And
brains, we are often told, differ among us, according to our gender or
sexuality.

Here, highly generalised social differences ('lesbians' versus 'heterosexual' women, for example), reflecting complex behaviour and beliefs,
is read onto the spaces of the body. Difference is mapped, and bodily
differences sought. Causation is then inferred – for example, the
supposition that, if lesbian brains are 'masculinised', then this must
have been caused by some aberration of hormones in the brain before
birth. (This is a new variant on an old theme: blame the maternal
environment, blame the mother.)

But what is crucial to this search for underlying differences is that
scientists must learn to 'read' the texts of the body's insides. Where
once difference was sought in the outer markers of the body – skin
colour, the size of external genitals – now it is sought inside. And it is
'inside' not only in terms of being under the skin, but is also a search

at the microscopic level: difference is now sought in cells or tissues, or molecules. But histologically prepared tissue must be interpreted; it is not simply a case of measuring length. Looking at Simon LeVay's photomicrographs of tissue from the hypothalamus for differences in the 'gay male' brain requires an act of faith: you have to learn to 'see' each area, to see the territorial boundaries. You have to learn to construct that visual space in parts.

IMAGINING INNER SPACE

What was read onto nature during the eighteenth and nineteenth centuries was not only the social order and a growing characterisation of different types of persons (see Foucault 1979): it was also mechanism, increasingly part of the way in which science offered both explanation and control over nature (Merchant 1982). As such interpretations of nature grew more widespread, so reductionist and mechanistic views of body organs prevailed; and as scientists relied more and more on the classification of nature, including disease, so the practices of anatomy changed to focus on pathologies and their classification. Thinking about the body itself as a machine became a significant part of the Industrial Revolution.[20]

Machines, moreover, make noise, as the seventeenth century scientist Robert Hooke pointed out. In a passage which illustrates the power of the mechanical metaphors that came later to dominate ideas about the body, he noted that bodily workings might be investigated,

> by the sound they make, who knows but that as in a Watch we may hear the beating of the Balance, and the running of the wheels ... and Multitudes of other noises; who knows, I say, but that it may be possible to discover the Motions of the Internal Parts of Bodies, whether Animal, Vegetable, or Mineral, by the sound they make, that one may discover the Works perform'd in the severals Offices and Shops of a Man's Body, and thereby discover what Instrument or Engine is out of order ... (Hooke, cited in Reiser 1978)

Sound, indeed, did become part of the diagnostic repertoire of doctors, albeit somewhat later. The invention of the stethoscope at the beginning of the nineteenth century enabled doctors to hear changes in the way that lungs or heart worked, or what 'Engine is out of order'. By doing so, they were imaginatively 'looking inside', trying to match the sounds heard to what they knew of pathology. They were also listening to a private conversation, to which the patients themselves were not privy.

Anatomists of the eighteenth and nineteenth centuries began to document various pathologies – the anatomical correlates of disease – within the corpses they opened. By the beginning of the nineteenth century, as a result, dissection had begun to transform physicians, according to one historical account, from 'word-oriented, theory-bound, scholastics to touch-oriented, observation-bound scientists' (Reiser 1978, p. 19). Now, tactile examination of the patient, to investigate parts of his or her body, began to dominate medical consultations; indeed, the interaction became much less of a consultation, as the opinion of the patients themselves became less important. What mattered now was the truth as revealed by the medical examination; in time, even this was to be replaced by the more 'objective' truths of the diagnostic machines.

But it was the development later that century of techniques of visualising internal structures that enabled physicians, at last, to look inside the living body, to bring it within their purview *through* the use of machines. Visualisation technologies are commonplace today. That development, combined with the growth of physiological experiment (on living animals) during the nineteenth century, opened the body's mysterious inner workings to the (male) medical gaze; it also helped to redefine boundaries – between sick and well, between normal and pathological.

The insides of living bodies first became more visible with the development of X-ray machines at the end of the nineteenth century. Not only did this seem to blur the boundaries between outside and inside, but it heralded a new era in which bodily insides came increasingly to be visualised by machines: internal functions thus came to be represented visually, in the form of visually displayed images (such as ultrasound scans, for example) or as graphic readout (the electrocardiograph, or ECG, for instance).

Examination of the body in clinical settings moved, historically, from observation of a person's general demeanour, through touch and hearing from the outside of the body, to visual and auditory pictures of the inside through technological production of images. These, as sociologist of science Bruno Latour has argued, have been critical in the development of clinical medicine as scientific. Scientists, he argues, are no longer looking 'at' three-dimensional nature directly, but at two-dimensional output from an array of machines. What persuades other scientists of the merit of their claims is precisely the graphs and diagrams: it is these products of what Latour calls inscription devices that matter if scientific claims are to be believed:

> No matter if people talk about quasars, gross national products, statistics on anthrax epizootic microbes, DNA or subparticle physics; the only way they can talk and not be undermined by counter-arguments as plausible as their own statements is if, and only if, they can make the things they say they are talking about fairly readable. (Latour 1983, p. 161)

Diagrammatic models do just that. The purpose of the models, graphs, tables and so on is thus to make arguments simpler, to make them more 'readable' – and hence to make them more persuasive. Whatever form of 'nature' scientists are studying, what they actually produce is a paper trace, a two-dimensional array.[21]

Visual representations in the form of diagrams and tables are thus critically part of the process of science; they are not simply perceptual aids, but are 'also a social process – an "assembly line" resulting in public access to new structures wrested out of obscurity or chaos. Instruments, graphic inscriptions, and interactional processes take the place of "mind" as the filter, serving to reduce phenomena of study into manageable data' (Lynch 1990, p. 156).

In Maria Trumpler's (1997) study of such images in neurophysiology, for example, she documents the scientists' reliance on symbols derived from electronics (based on their own experience of having to build the electronic equipment from scratch). This in turn has structured how future generations of scientists have thought about nerve function, as though nerve membranes almost literally incorporated the little symbols of the electrical circuit. So insistent is the message that I still invoke mental images of circuit diagrams whenever I imagine a nerve cell. I still think how to 'short-circuit' the action of the nerve. It is almost as though there is a miniature circuit board in the nerve cell membrane, rather like the tiny homunculus that early microscopists claimed to see in the head of a sperm.

Crucial to these analyses of scientists' use of visual models is the acknowledgement that scientists (and others) must learn to *read* the images, just as students of film studies learn to read such cues as lighting. Even something as basic in scientific training as using a microscope requires 'learning how to see'. I have been struck many times by the bewilderment many students show when they peer down a microscope; as Evelyn Fox Keller has noted, most of us see precious little, except our own eyeballs when we look down through the instrument (Keller 1996, p. 110). And it is not that students cannot see anything, once they have the slide in focus; it is rather that they do not know *how* to see, how to interpret and frame whatever lines and colours fill the field of view, how to distinguish the 'real' from the artefact, the cell from the air bubble.

Diagrams and models must also be read. One assumption underlying their use in scientific papers, for example, is that they do not represent a particular instance (this sodium channel through the membrane, for example, or that subcellular structure), but they are understood to represent something universal (Lynch 1990) – even when they are placed next to an image of something specific (a photograph, say). The photograph may be *of* a particular cell or tissue, but the abstracted line diagram is the generalised cell/tissue.

It is that step, from the particular to the universal, that students must learn to incorporate into their 'reading'. No-one could easily draw an exact copy of the specific tissues under a microscope. They must learn to abstract, to render the image into a universal form, just as trainee doctors must learn to 'read' the tissues of the dissected body (Good 1997). I remember well my school lessons in histology; I was required to learn to abstract, and to compare 'my' tissues to diagrammatic images in textbooks. Thus could I learn how to abstract visual images into an accepted 'universal' form. Identifying particular cell types for an examination then required that I learned to match *whatever* I saw in front of me to a remembered idealised image. From there, I would have to recognise and give a name to whatever I saw, or to reproduce that idealised image from my memory, as a drawing.

But does all this rather theoretical stuff about how scientists rely on certain forms of visual representation matter to ordinary folk? Does it matter to how we think about bodily interiors? Perhaps the electrical circuit diagrams are familiar only to a few; most people would not, I assume, readily think of circuit diagrams when, or if, they think about how nervous systems work. Nor did I readily extrapolate from 'my' lung tissue or ileum as seen in the circular field of the microscope to *my* own lungs or guts.

Some of these inscription devices have great persuasive power outside the laboratory. Think for a moment of the story familiar to anyone who has ever watched hospital dramas on television – the cardiac arrest. You know what happens: people rush about, electric shock pads are applied to the patient's chest, intravenous drugs are given, someone tries cardiac massage. Meanwhile, the oscilloscope screen bleakly tells the story: despite the heroic efforts, the team has failed. We all know how to interpret the still line that crosses the screen. The machine in this familiar narrative stands in for any understanding of what hearts might look like, or do. Its printout may not be fully decipherable by most people, with all those blips and sounds. But we know what the shift from blips to stillness means. The lines on the screen become the arresting heart. We do not need to be persuaded through, say, dramatic effects, that the person is dead: the graphical image powerfully convinces.

Even in written texts that aim to popularise scientific ideas, readers must usually do some work to interpret graphical or diagrammatic images. We have to understand how, say, a photograph of some phenomenon in nature relates to the scientist's graphic display; we must learn to translate one into the other. It is, argues Greg Myers in relation to his examination of biological texts, that very work of translation which makes the story the images tell so powerful (Myers 1990, p. 262).

In the case of visualisation technologies in a hospital, however, it is not only other scientists/doctors who must interpret the machine output; it is also (partly) the patient. But the patient (and her allies, her friends and family) have not usually learned to 'read' the output in the way the 'experts' have. The technology thus profoundly structures the social relationships of medical personnel and lay people.

Moreover, the technology also serves to define when we are sick or well. A routine scan might reveal something untoward – perhaps a shadow in some internal organ. We can thus, at a stroke, be moved from the category of being well (and we probably felt perfectly well when we entered the hospital) to being defined as 'sick' – bearer of a potential lump. In this case, patients are not likely to be shown the visual output, the shadow on the scan: the technology determines who can have the knowledge and how it is to be understood (as well as when it is understood: photographic products from X-rays or scans have to interpreted by the radiographer before the patient is told anything, at the next appointment).

How images are read has considerable implications in medical education, in structuring the knowledge students gain. In his study of what goes on in medical training, Byron Good noticed a typical pattern of teaching: a slide showing epidemiology of a disease would be followed by a picture of a patient, followed by a slide of a pathological specimen, then tissue samples from the specimen, cells from the samples and so on in a seemingly logical progression (Good 1997).

The reductionist message is clear, argues Good: 'Surface phenomena of signs, symptoms and experience are shown to be understandable with reference to underlying mechanisms at an ontologically prior level. Even broadly incorporative biopsychosocial models, articulated in the language of systems theory, represent biology at the center, social relations outward at the periphery' (ibid., pp. 75-6). It is, perhaps, little wonder that doctors and patients rarely seem to speak the same language.

I have dwelt on visual images, machine readouts, in science, partly because of their rhetorical significance, and because of their practical importance at the hospital bedside. Images persuade. In that sense, they can act to persuade other scientists or doctors, and help to give further authority to the scientific voice.

Scientific stories of what there is inside the body have moved, then, from the examination of the location of different organs and tissues in the dead body, through examination of external signs of pathologies, to the readout of devices which 'tell' us about what is 'really' happening inside. The internal body is, through these processes of diagnosis, normalised and its inner workings mapped. The signs that can be read indicate normal versus pathological, healthy versus sick. In these shifts, moreover, any knowledge that the patient might have about her own insides has become invalidated; what now defines sickness *is* the readout from the machines, in whatever form – photograph or graphic display.

These scientific stories of the body as mechanism, as a system of separable parts, have become triumphant. It is part of our culture; we no longer believe in the fluxes and flows of earlier epochs. Or do we? No doubt most people in Western culture would tell the same stories of the inside bits of the body. But of course they would do so if asked about 'what is inside the body': the question almost presupposes a 'scientific' answer.

But other perceptions and representations of the inner body may still circulate in our culture; it is just that we are unlikely to call upon them in the context of scientific explanation. These other representations are, in effect, short-circuited by the power of the story of the body–machine. I suspect, however, that earlier representations of the inner body as flows, as space, have echoes in the cultural imagination; it is not the body in mechanistic parts, but the fluidity of the body that is invoked for example in poetry or stories of sexual desire. These are a far cry from the highly structured images that we might encounter in the textbooks of science. Diagrams, in particular, often conceptualise our insides in terms of boxes – far removed indeed from the poetic imagination, which can fill our inner spaces with imagined things, like furnishings in a room:

> with her   in
> side is very much like out
> side   only with
> furniture in it

(from 'Leaves and Potatoes', Yeo, 1988)

4

Spaces and Solidities:
Representing Inner Processes

The corpse, lying on the dissecting table, is pure spectacle. It is
the body exposed in its isolation to the full light of objective
consciousness: a blinding, antiseptic whiteness; a body of purity.
(Romanyshyn 1989, p. 17)

From dissection to reading graphic displays from machine – the
story I have just sketched is a tale of visualisation. The dark
recesses of the body are brought into the light; the inner spaces
become continuous with external space. As they do so, they become
images available to all of us. In this chapter, I want to turn from that
overview, to focus on a particular kind of representation or visualisation of the inside of the body – that provided by scientific diagrams.
The abstraction of these diagrams from the reality of blood/guts and
internal organs is embedded in the shifts of perception I outlined in
the last chapter. Abstraction is familiar; we all know what those line
diagrams in elementary textbooks mean. Of course we know that real
uteruses don't actually *look* like that. Don't we?

Those abstracted images, however, conceal their cultural origins. I
commented earlier on how even feminist students often suspend their
critical faculties when confronted with 'scientific' diagrams, something
they would never do with other visual images. Here, I want to look
more carefully at some of these images, drawn from elementary textbooks, and then go on to consider newer forms of abstraction, on the
Internet. What does such abstraction do to our perceptions of our
embodied selves? And what are the implications for the gendered body?

Abstraction in these two-dimensional visual representations is now so much a part of science education that we can easily overlook it – and what it obscures.[22] The images of dissected human parts in Vesalius' sixteenth century work are very different from many of the images of bodily insides that typify elementary biology textbooks today; not only do they have finely drawn detail, but Vesalius' drawings are set in an environment, replete with details (including the gallows!). This is very different from the kind of anatomy that most of us learn from school textbooks.

Anatomical drawing gradually became more abstracted. From the highly detailed drawings of those early anatomists, complete with shading and against a suitable background, anatomical drawing became more and more removed. Some detail (shading or use of lines to emphasise shape) may remain, but a large part of the drawing has now become highly diagrammatic (see Laqueur 1990; pp. 163–8, for illustrative examples).

There are many examples. Many of us could roughly reproduce a line diagram of, say, the uterus – even if we would be unable to draw a detailed drawing. One example I have come across (from a book published in the 1980s) shows a baby at a woman's breast. The baby is drawn 'artistically', with details and shading to convey realism; it is looking up at the mother and puckering its mouth around her nipple. But the diagram is intended to illustrate how that sucking can stimulate the mother's hormones. So, the mother's head shows few details apart from the outline only. Inside the space of that outline are lines and arrows representing the hormonal control of milk output via her brain. Arrows go to her breast, there to meet with the shading around the nipple where it joins the baby's mouth.

Another example comes from a century earlier, in a gynaecological textbook. Here, the doctor is drawn, in suitably professional attire as he attends to the woman lying down. She is shown fully clothed, in dress of the period – at least from the waist up. This much is all detailed with shading. But below her waist is simply a line diagram of her uterus – and nothing else – apparently floating in space a few inches above the bed and a few inches from the doctors hands.[23] The diagram/drawing was intended to show how doctors deal with a prolapsed uterus.

Students to whom I have shown these images usually find them disturbing, especially the second (although they laugh at it, too). We may be familiar enough with abstraction in scientific diagrams as we learned them in simplified form at school. But those are images abstracted *from* more realistic representations – indeed, from the bodies that they are supposed to represent. It is precisely the juxtaposition of

Elevation of the uterus in prolapse. The physician replaces and lifts the uterus ...

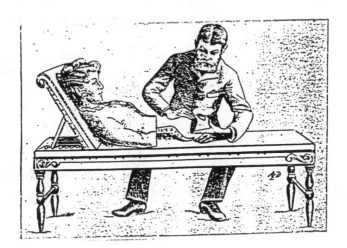

... while the assistant reaches into the pelvis to catch it.

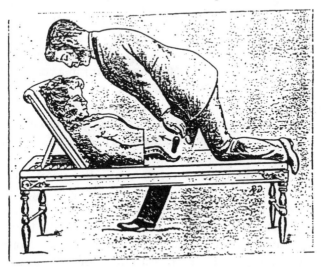

Figure 4:1 Nineteenth-century illustrations, to show how a doctor might treat a prolapsed uterus (from Dickinson 1892, p. 775)

highly stylised images with more finely drawn and artistic renderings that remind us of what is represented. The contrast is stark. Next to 'realistic' components, the abstractions evoke a sense of bodies in parts, an echo of dismemberment.

Scientific diagrams were so much a part of my own education that it took some effort to look again at them, but this time with a more critical eye. In doing so, I went back to a number of elementary text-books (from approximately 25–30 years ago). Some of these were ones I had to learn from; others were books I had used when I first began teaching. How abstracted, I wanted to know, were the illustrations that students – including me – were expected to learn? The first book I examined was a very elementary human biology text, *A First Course in Hygiene* (Lyster 1966). This has 165 diagrams, approximately half of which illustrate parts of the systems of the human body. What first struck me here was how very few of the organs/systems described were illustrated *in situ* – or at least, not in a whole body outline. Those few that even put them into an outline showed only the thorax.

All the illustrations were diagrammatic – that is, they were abstract images, with relatively little use of shading or attempts at realism – with the exception of drawings of what to do in artificial respiration. Here, the drawings were of people (white males), rather than highly stylised diagrams.

Shading is sparse in these line diagrams. Some appears on drawings of the skeleton, or of the heart. Interestingly, the strongest use of shading was in two diagrams of sections through the human head (again, a head drawn to represent white males). Here, the organs are shown close together, so shading helps to embed them in a context. But these are the only illustrations that do so.

A second text I reexamined was an *Introduction to Biology* (Mackean 1965). Here, I found similarly abstracted diagrams (usually to illustrate human organs, or a generalised mammal); but some of these were juxtaposed to photographs taken from a dissected rabbit (photographs show the alimentary canal, the heart, and the female reproductive organs of the rabbit). To illustrate particular systems or organs, the organs are usually sketched in outline and stand alone, devoid of context.

In the next few pages, there are further photographs to show 'Juno, the Transparent Woman' – similar to the plastic woman of my child-hood. Juno's transparent skin reveals more of the inner detail, the network of veins and arteries, the white fat cells under the skin of her breasts. Later still, another 'transparent woman' appears. This one is pregnant, and her transparency reveals nothing inside except the late-pregnancy foetus. In nearly all these texts, women's anatomy is contrasted to that of males, and is often portrayed with a pregnant uterus. Women's bodies, as so often, are intelligible inside only if gravid.

Higher level texts followed a similar pattern, except that here, information in the form of graphs and tables begins to appear, and even to predominate: thus, a graph showing how oxygen dissociates from haemoglobin at different concentrations may appear in a chapter on blood and circulation.[24] Graphs and numerical labels may appear embedded within a line diagram, or indicated by arrows to show how they relate to body parts represented by the diagram. Graphs now come to predominate.

The chapter on the thyroid gland, for example, in Ganong's *Review of Medical Physiology* (1973) has four diagrams (black and white drawings) to illustrate actual tissues or cells from the thyroid. The text includes three photographs, of patients with thyroid disease. The remaining eleven figures are of tables and graphs showing, for instance, what happens to an animal's metabolism after injection of the thyroid hormone thyroxine.

The stark black and white drawings of these older books are partly replaced in more recent texts by colour. Later (1980s–1990s) elementary biology books for example continue with the abstract diagrams, though now shading as well as colour are more common. Accompanying photographs add realism; they might show a child visiting the dentist, or a glistening photo of a fleshy valve removed from a human heart. Cartoons also feature, adding humour. Illustrations, too, are representative of more than just white males; both doctor and patient can be black or white, male or female (see, for example, Beckett and Gallagher 1989).

What I found striking, particularly in the earlier textbooks, was the way in which diagrams of organs or systems quite rarely used context. Rather, the organ was represented as somehow free-floating on the page; or perhaps it was floating between two wavy lines symbolising the torso. The reader is supposed to infer the precise location within the body (or at least within the *diagrammatic* body; whether readers necessarily extrapolate to their own bodies – with organs in mirror image to the diagram – is another question). Sometimes, the outline is highly stylised, resembling no known organism; in this case, it might, for example, stand for a 'typical vertebrate' or a 'typical mammal'. Thus the typical vertebrate of Romer's (1970) *The Vertebrate Body* is a rather tube-like affair, closer to some kinds of 'primitive' fish than to more complex vertebrates.[25]

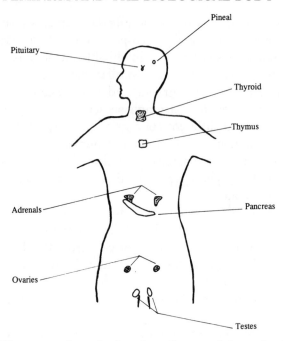

Figure 4:2 Diagram to show the location of some of the endocrine glands. This is a typical androgynous diagram (drawn as a composite from several textbook illustrations) showing both ovaries and testes, but always with an outline of 'man'.

If a 'whole body' outline did enclose an organ in the diagrams I examined in elementary books, the outline was almost always that of a non-human animal – a dog or a rabbit, for instance. A clearly human outline appeared largely in illustrations of the head or brain, or in the more advanced texts. The human head, too, was further prioritised by the use of fine detail and shading. Dualist assumptions – human over animal, mind (head) over body – pervade illustrations just as they do the written text.

Dualisms of gender can, however, be overridden in the service of scientific abstraction. One example, from a book focusing on hormones (Donovan 1988), is an illustration to show 'The location of certain endocrine organs in man' (p. 10). The usual outline is there, which is certainly 'man' in its shape. Inside, are a few of the usual organs floating in space – the pituitary, the parathyroids, the thymus, the pancreas, the adrenals. The illustration also includes both the testes (which add to the shape of the overall body outline) and, floating just above them, the ovaries – the androgynous diagram.

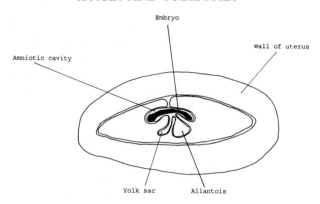

Figure 4:3 Simplified diagram to represent pregnancy and the relationship
between the embryo/foetus and surrounding membranes in a 'typical'
mammal. This is a composite, redrawn from several textbooks. Where species
is specified, the diagram captions typically refer to gestation in 'man'.

The use of space in diagrams is sometimes itself gendered.
Diagrams, for example, of the female reproductive tract often portray
the uterus *as* a space, by contrast to the solidity of the male equivalent.
In one elementary textbook (Beckett and Gallagher 1989), for example,
the diagram of the uterus shows it in the middle of the page, abstracted
from the sketch of a (white) female torso. The accompanying text notes
that 'The womb or uterus is a bag in which a fertilised ovum develops
into a baby' (p. 149). The male tract, by contrast, is illustrated in its
place in the body and drawn to emphasise the solidity of the penis.
Women's anatomy *is* space, disappearing to become merely a
receptacle for the baby, while men, it would seem, are firm and solid.

Now stylising and simplifying for the purposes of diagrams serves a
purpose, particularly in teaching – it can get the message across. I
would certainly not wish to use details of shading and so forth when
teaching elementary anatomy and physiology, but would need to use
such simplified teaching aids. Using details and context every time we
wanted to illustrate, say, organs in the abdominal cavity, would be very
tedious. What I want to emphasise here, however, is how we might
read these stylised diagrams and what is omitted (or not) from them.
In order to simplify, the scientist or artist must make choices about
what is included, what is shaded, how the structures are to be represented in stylised form. It is this choice which can influence the
reading – which in turn can become a *gendered* reading.

What is lost in the processes of abstraction? First, something is lost in the translation of a three-dimensional structure onto two-dimensional space – the initial drawing. Then details are lost which convey shape and form – the shading. Finally, all details are lost of the context of the organ within the body, or of the body itself; the organ/body becomes free-floating. The body is presented as if sectioned longitudinally; the front is removed so that we can 'see in' to the outlined organs. Now, the only details that might be added are notations that indicate that what we are seeing is a *scientific* text; these notations might include mathematical symbols, arrows, graphs, tables superimposed on or juxtaposed to, the diagrammatic organ.

The abstractions of the diagrams, moreover, work closely with the abstractions of the written text. To take one example, from one of my textbooks on the molecular biology of steroid hormones (Schulster *et al.* 1976). Here, the authors note the problems of measuring hormone levels in the living body, including the problems posed by relatively inaccessible organs. The solution, they tell us, is to transplant the organ 'to an accessible place'. The diagram (p. 37) shows the outline of a sheep's head and neck. The arrows and labels tell us that an ovary and its artery are found there, transplanted surgically. 'This preparation', the text notes, 'allows direct access to both the arterial and venous sides of the ovarian circulation' (ibid., p. 36).

Abstraction here exteriorises the organ in question – not literally (except temporarily; it was linked up in the experiment to the carotid artery), but figuratively. The diagram brings it to the surface, exposing it; its existence as an organ out of place seems to make the exposure more explicit. The language, as usual for science, also obscures; the agency of the scientist is omitted, while 'this preparation' reminds us that living animals can be literally rearranged, prepared, to suit the experimenter's needs. The language works closely with the diagram to create abstraction.

Abstraction, furthermore, enables the illustrator to omit details, to censor what goes into the image. The commonest way diagrammatically to represent the 'female reproductive tract' is from the front (at least in elementary textbooks). The uterus can then be drawn as a triangular shape, with Fallopian tubes either side. Below are the two lines representing the vagina. This image necessarily omits details that are scientifically irrelevant to the narrative of reproduction: the clitoris, for example, does not appear in these highly stylised and abstracted diagrams. It is removed by the longitudinal section, which has erased everything situated in front of the uterus.

In theory, diagrams representing a section cut from the side should be better; after all, this should 'cut' through the whole of the external genitals. But no. The clitoris seems to escape detection. Even feminists have contributed to the erasure of the clitoris, which has received relatively little scholarly attention (Moore and Clarke 1995) – an absence which has been dubbed 'critical clitoridectomy' (Bennett 1993, p. 242). Analysing medical texts, Moore and Clarke note how the labelling of the clitoris was actively removed from the 1948 edition of *Gray's Anatomy*; its absence was repeated in several other texts through the 1950s and 1960s. Note that they were examining medical texts which tend to illustrate female anatomy from the side. There is, then, no reason why the clitoris should not be labelled.

Despite the detailed labelling that inevitably was included in feminist books about women's bodies (such as the Boston Women's Health Collective's *Our Bodies, Ourselves*), there has been what might be called a subsequent clitoral backlash, another erasure. Recent texts, Moore and Clarke found, tend to ignore or downplay the clitoris, and to emphasise the narrative of reproduction.[26] The vagina exists for the penis, and lubrication exists for heterosexual intercourse in these narratives. These forms of abstraction – in the diagrams and in the accompanying text – clearly have implications for women. We are divorced from our bodies; even those parts that give us pleasure disappear or have no names.

These processes of abstraction create spaces where 'something used to be' (Figlio 1996). Viewers/readers must do work to fill in the gaps, to give the image meaning. (And in the case of the missing clitoris, the gender of the viewer/reader becomes highly significant.) But the meaning as constructed in medical diagrams is not arbitrary. We are not supposed, as viewers, to see the thoracic cavity as, say, a cooking pot (as, perhaps, the Medieval surgeon described in Pouchelle's (1990) study might have done). Rather, we are to read it as scientific, as mathematical, as abstract.

LEARNING TO READ THE IMAGE

Yet what of the impact upon the reader – or the potential patient? Some readers may need to use the knowledge directly, for example, in medical practice. Surgeons, of course, must do more than simply the intellectual work of filling in the gaps left by abstractions; they must act upon their knowledge. A surgeon equipped only with the sparse diagrams of the elementary books would undoubtedly kill the patient.

Anatomical atlases, from which medical students begin to learn the finer details of human anatomy, begin to return from the abstract diagrams of the school science books to the realism of earlier centuries. Doctors, and especially surgeons, need to know what the organs 'really' look like before they can begin even to dissect a corpse.

In his study of surgeons and surgery, Stefan Hirschauer (1991) suggests that the practices of surgery themselves help to recreate the images drawn from the anatomy books. 'Dissection aims to present organs in the isolating style of the anatomical atlas', he argues, 'The drawings show neatly separated organs; in the patient-body this state must first be produced by isolating them with the knife ... to a layperson, this procedure increasingly disfigures "the body" – as it is known from everyday life – for the surgeon [this] exposition *creates* "the body" – as it is known from the anatomy book' (ibid., p. 301).

Hirschauer goes on to argue that the low level of abstraction in anatomical texts serves to ground the linkage between the 'anatomical *knowing that* of the visible, and the anatomical *knowing how* of making something visible' (ibid., p. 310; all emphases in original). The surgeon must know how to make visible, just as the student of elementary biology must know how to see. He or she must learn to read, to interpret the stylised and abstracted images of textbooks; later, in surgical practice, they make the body conform to the images learned so well.

The processes of abstraction carry several, interlocking, messages. First, as Michael Lynch has observed (Lynch 1990), scientific diagrams accompanying photographs are not stylised just to simplify (though simplification is one effect, which helps to make the image persuasive): rather, he argues, the line diagrams are *idealised* and normative. In reading the juxtaposition of photograph and line drawing, we move from photo (itself an abstraction from the original material) *to* diagram. The diagram becomes, argues Lynch, a mathematical icon, a 'theoretical domain of pure structure and universal laws which a Galilean science treats as the foundation of order in the sensory world' (ibid., pp. 162–3).

Students of biology, at whatever level, must learn to read the abstracted images. At higher levels, the integration of line diagrams with mathematical representations is much more explicit: here, graphs and even formal equations predominate. This translation changes the shaded, topological surfaces of anatomical drawing into rectangular, usually two-dimensional, surfaces: the inner surfaces of the body become geometric. Now the reader has to learn to interpret these graphs, to read them in relation to those line drawings that she learned at an earlier stage. The abstraction from the flesh of the body is now well underway: even the line drawings of bodily organs disappear into graphical output.

And this is where the formal language of science seems almost to be necessary, to speak of the properties of *the* blood or *the* lung. Thus I could learn to reproduce and interpret the graphs explaining the chemical properties of the blood, I could draw the histological struc-ture of the lung tissue and relate it to graphs explaining the transfer of gases in and out of the alveoli. But only rarely did I ever wonder what was happening to *my* blood, to *my* lungs. My lungs only made their presence felt when I endured the pain of pleurisy or when I had a bad cough.

As I have noted, instruments and graphs in the laboratory them-selves help to organise how the knowledge is produced – how, that is, the object under study becomes constituted as a scientific object. Lynch suggests that: 'The details of laboratory work, and of the visible products of such work, are largely organised around the practical task of constituting and "framing" a phenomenon so that it *can* be measured and mathematically described. The work of constituting a measurable phenomenon is not entirely separate from the work of measurement itself' (ibid., p. 170; emphasis in original).

Put another way, the organisation of the physiology labs and the way that they utilise measurement devices, is a critical part of the process whereby increasingly abstracted images of 'how bodies work' are generated. The same is true of hospital wards, except that there it is patients' and their bodies who are marginalised by the abstractions. By the late twentieth century, even lay understandings of bodily insides are shaped by that history of laboratory measurement and graphical output. Even the most elementary of the texts I examined uses graphs and mathematical tables. The body's insides have *become* numerical. It is a very long way from the ebbs and flows of humoural doctrine.

The geometricised, numerical, body is a body separated into parts, each carefully labelled. This, argues Robert Romanyshyn, writing about art and technology, is part of the heritage of the discovery of linear perspective, which changed not only art, but also human perception of ourselves and our bodies. Developing linear perspective meant splitting our vision of the world up into fragments, and distan-cing ourselves from it. 'This anatomising vision', suggests Romanyshyn, 'turned toward the human body, gives rise first to the anatomical body, the body fragmented into parts, the corpse; and second to the body as reflex movement' (Romanyshyn 1989, p. 77).[27]

While I can trace the persistent abstractions over the last few decades of physiology texts, there are also new forms. Digital imaging and ultrasound, for instance, allows us to 'see' inside the body without even lifting a scalpel. But now abstraction has proceeded further still. Ultrasound images, for example, are notoriously difficult for anyone to read unless they have been trained to do so. All that displays on the screen is a patchwork of black and white, that (we are told) corresponds to this or that internal organ. These images bear little resemblance to the line drawings of the textbooks.

Yet they must be interpreted. For women, one of the most salient images comes from ultrasonography in pregnancy, showing the foetus (or 'your baby', as it is usually described by the radiographer). What the image shows – how it is read – is a deeply cultural act. In a cross-cultural study of the experiences of women undergoing ultrasound examination in Canada and Greece, Lisa Mitchell and Eugenia Georges noted how differently these women understood the images they saw on the screen (Mitchell and Georges 1997). Women in North America, they argued, were more likely to see the image (the baby) as separate from themselves, while Greek women tended to see the foetus as in relation to themselves. They point out that this difference is at least partly related to the ways in which pregnancy is experienced; in Canada, pregnancy is more medicalised, and women more likely to read medical books and pamphlets which construct the foetus as having personhood (enough, indeed, to create it as a medical patient in its own right).

Images of the foetus are, moreover, extrapolated *from* their context – the uterus inside a living woman. As Ros Petchesky has cogently argued, such images create a sense of the foetus as a 'spaceship traveller', floating free (Petchesky 1987). The politics of such portrayal, with regard to abortion and to issues of foetal personhood and women's reproductive rights, have been much discussed in feminist writing (see also Duden 1993). What those images also do is to recreate a sense of internal space. The spaceship traveller is not entering some galaxy of outer space, but is moving through the inner 'space' created by such images.

Other forms of imaging can recreate space to greater or lesser degrees. Doctors might subject a living patient to a scan (such as an MRI scan – Magnetic Resonance Imaging), in which a signal (similar to radio waves) is passed through the patient's body. A computer analyses these signals and builds up a three-dimensional picture of the person's insides. Like other forms of imaging, these pictures need to be analysed; they have to be read in order to reach a diagnosis. We have to know how to read some of those inner spaces, for example (such as the spaces containing the cerebrospinal fluid). But in many ways these images are less abstract; some of the pictures the MRI scan generates look a little like, well, meat.

SURFING FLESH: BODILY INSIDES

Scanned images are now on the Internet. In particular, the Visible Human Project offers any net user access to see inside a human body. Discovering it for the first time while 'surfing the web', I was reminded of the 'Visible Woman' model of my childhood. Inside the box were organs in pieces that I had to assemble, and then to paint in different colours. I then had to place these inside the transparent outer layer. This woman was, of course, young and white (insofar as that can be judged from her facial features). But most significantly, she was of reproductive age: I could perform a plastic hysterectomy on her, removing her uterus and replacing it with a highly pregnant one.

The Visible Human Project has been set up through the US National Library of Medicine (part of the National Institutes of Health). It began with the (dead) body of a white male; later a female body was added. Reminiscent of earlier use of corpses from the gallows, the male body was that of a man who had been sentenced to death by lethal injection. The woman, a 59-year-old, had died in her sleep. Their bodies were photographed and scanned into a massive database, accessible from the Internet. The images, taken from thousands of scans of these human bodies, can be selected or rotated, even animated, by the user. We, the viewers, can exert control over the pictures of flesh that we rarely could before such computerisation. We can bring the inside out, and bring it before our gaze.

Evelyn Fox Keller, in her examination of the 'biological gaze', has noted how it may once have been possible to think of the eye and vision as passively observing nature (Keller 1996). But the development of technologies such as microscopes contributed to a shift in the predominance of the naked eye. Now, it is possible manually to insert pieces of a cell or of DNA into other tissues, while watching what one is doing under a (more powerful) microscope. Seeing and touching have merged, she argues, in the practices of modern biology.[28]

We may be able to gaze at, or into, the bodies that are now 'accessible' on the Internet, even if we cannot touch them. But, to see them on the screen, we have to carry out a series of physical manoeuvres –

clicking the mouse this way and that. The merging of seeing and touching to which Keller refers thus takes on a different form. The large scale movements of the dissecting physician have become replaced at the moment of our observation by the remote movements of the scanner, and thence of our hands on the mouse or keyboard. The organs of the body become at once closer (*we* can 'touch' them electronically, slice through them; we do not have to rely on 'them', the expert surgeons, to know bodily insides), yet distant (these bodily insides now exist only in cyberspace). At least we can keep our hands clean.

The technologies of visualisation thus become part of the social processes by which we come to understand what it means to 'see' inside the body. Think, for example, of how we might learn to 'see' the images derived from the ultrasound scan. The very process of seeing entails a mutual arrangement of bodies and their component tissues, as well as of the technologies themselves. The internal spaces are thus made visible partly because of the way that human bodies – the participants in this drama of visualisation – are arranged in sociotechnical space. In the case of bodies made visible through cyberspace, that arrangement is even further removed from the act of gazing.

At first glance, the Visible Human Project seems to be somewhat less sexist than many textbooks; it *does* include a woman (although she was added later). Yet, as Lisa Cartwright (1998) points out, cultural narratives about these bodies tend to ascribe them to 'conventional heterosexual family models', echoes of the myth of Adam and Eve. Cartwright speaks of them as the (male) 'Internet Angel and [the] Menopausal Housewife'; the woman's body is inadequate because of her menopausal status, and another, younger, body is being sought by the Project.

The narrative of woman's body as reproduction is further, and disturbingly, echoed by the analogous dataset produced from the University of California at Stanford and reproduced on their webpages. Here, you can find the Stanford Visible Female – the cryopreserved pelvis of a younger woman. A 'suitable cadaver', the webpages tell us, was found in 1993, a woman of reproductive age. Or rather, the 'specimen' was of a 32-year-old female. But it is not her whole body that is on the Internet; it is just her pelvis. The text is accompanied by photographs showing the (male) researchers with what looks like a large lump of meat encased in ice, and being cut with a bandsaw. This object is the sectioned pelvis, from which the rest of the woman's body has been cut away.[29] Finding these images, I experienced the same kind of disturbance evoked by the images I described earlier, from the

76

nineteenth century, which juxtaposed the apparent realism of a shaded line drawing with the stylised diagram of a uterus. What matters is not the whole woman; in the 'Stanford woman', it is just her pelvis, abstracted from its context. The imagery, the language (the typical passive voice of the scientific account) – all these things remove us, distance us from whoever that woman might have been.

I am well aware of the potential usefulness of these digitised images; they undoubtedly can help to further knowledge and to enable medical students better to understand 'pelvic anatomy' (synonymous with female pelvises). And I am also well aware of the need to preserve such 'specimens' in order to render them permanently visible. Preserving in ice has long been a technique used in electron microscopy, for example.

None of that helps the disquiet I feel on seeing those images, or reading the accompanying text. The passive voice very effectively obscures what goes on in scientific practice (see Gross 1990; and, in another context, Birke and Smith 1994); much is omitted from the narratives. Consider the webpage text: 'The cadaver was transected just above the level of the iliac crests and at the upper thighs …The specimen was then placed in a [box] … A liquid mixture was poured into the peritoneal cavity … [and then frozen] … The container hold-ing the frozen specimen was initially sectioned in halves with a band saw … [and then photographed] … Sectioning was continued until the anatomic region of interest was completed' (Stanford Visible Female webpage: Project Overview).

Notice how the woman herself has been progressively erased – first, by her death and conversion to a cadaver. Then, she is erased further by being 'transected' (somehow, saying cut in half would not be appropriate). Once transected, she becomes a 'specimen', into whom (or which) liquids can be 'poured': she is now a preparation, a receptacle (further contributing to the notion of space). Next, she is sectioned repeatedly until the parts wanted by the scientist are revealed. Even her disembodied pelvis can be raped.

The Stanford webpages lead to a series of three-dimensional model images derived from the subsequent data. Here, we can view the 'Uterus, tubes, ovaries and the lower urinary tract *nestled* in the pelvic floor muscles', or the 'Vulva with urethra, vagina and rectum *entering* posteriorly' (my emphases). This is fairly explicit sexual imagery; no wonder rape flashed through my mind. It is, moreover, a double assault. Not only is the construction of this 'specimen' described in terms redolent of sexual assault, but so too is the reader/viewer a participant. And that includes me, as viewer of these images.

The Visible Human Project is a little less disturbing. Here, we can learn how to 'March Through the Visible Man' (or woman). We can proceed to 'Flying Through the Body', where we can undertake a 'flight through the bone surfaces', or 'fly around the thighs' (another phrase with sexual overtones). Or, we can even 'Make Your Own Visible Woman', using visualisation techniques. Linguistic erasure is here, too. We can read how researchers applied 'a single [computerised] methodology to extract skin, bone, muscle and the bowels from the fresh CT data' (quotes from webpage 1996). They are not, of course, literally extracting bowels, but the digitised data corresponding to the scanned images of the bowels.

The photographic images of the bodies contributing to the Visible Human Project suggest that both bodies were white (Cartwright 1998). These – and the vast majority of other images for visualising anatomy that are available on the Internet – have a pinkish-beige colour (called 'flesh' by some crayon manufacturers, as Cartwright notes). Darker colours enter the spaces of images where muscle – real flesh – occurs. There is, as Cartwright observes, great potential for highly racialised readings of these images.

Where once it was doctors/scientists who could probe the 'dark secrets' of the body, like colonial explorers, now it is open to all of us (rather like the neocolonial practices of exploration, now open to anyone who can afford the airfare to exotic locations). Anyone who can use a mouse can enter within. But, despite the narrative of flying or marching through these bodies, the images are rather difficult to read. Part of the problem has to do with the inbuilt incommensurability of text and images in hypertext. It is not currently possible, therefore, to label the images, so creating what Lisa Cartwright has called a 'cataloger's nightmare', a plethora of images without any guide to what goes where. At least my childhood 'visible woman' model had a chart.

Cartwright also notes that the very detail of the digitised images renders them difficult to use by the many, disparate, users of the project. They thus become less universal, while appearing to represent us all. Moreover, to read the images requires that the viewer takes the mental step of converting one to another. It is not easy to make the leap of imagination from the two-dimensional slice to the whole structure. They seem to bear little relationship to how we might imagine organs in the body. The three-dimensional models derived from the Stanford project are also abstracted; they look like models made from papier mâché, coloured with crayons to make them look pretty.

In some ways, these representations of internal organs seem to return us to the more realistic interpretations of the early anatomists. Here, we have shading, and colour. With appropriate computer software, we can rotate the images in three-dimensional space. Yet these images are also more abstracted, in the sense that the shapes and lines we can bring up on the screen no longer seem to symbolise clearly delineated structures (except in the papier mâché form of the Stanford interpretations). Rather, we now have blendings of light and shade which, we are told, are a slice through the thoracic midline. This bit of colour is part of the heart, that one part of the lung. Interpreting those simple lines of the elementary textbooks was a much simpler task.

ILLUMINATIONS: SPACE AND IMAGES OF BODY ORGANS

From the fine detail of the drawings of the early anatomists, which began from illustrations of the whole body and took the viewer progressively through, following the path of the scalpel, there has been a move towards diagrammatic renderings, towards abstraction. First, the organs might be illustrated in a partial context – the thorax, for instance. Then, they might become illustrated as separate organs, devoid of context. Yet they still have solidity, for at least some shading helps to give them shape.

The next stage of abstraction is the removal of shading. Now, the organ becomes more stylised, and increasingly removed from context. When shading is present, we can more easily make the imaginative leap to understanding how the organ might look in a body. Once the organ is represented only by lines, however, that leap is more difficult; it is much harder to reconnect to the idea of a body when that body has disappeared into the empty space of a blank page.

The next stage is to represent the organ more diagrammatically. Thus, the heart might become a set of four boxes, rather than showing its shape and contours. And finally, the organ or organ systems are reduced to a series of geometric shapes, connected with lines and arrows. By following the arrows, we can learn how the blood flows from the ventricles into other parts of the body. Now, computers allow another form of abstraction, the construction of three-dimensional images regenerated from diagrams. Prettily coloured these may be, but they remain abstracted from any context. Rather, they rotate eerily on the screen, disembodied vulvas, vaginas, uteruses.

Think, for example, of the line diagram of a uterus; this may (or may not) have a line drawing around it representing a woman (probably no head, but curves in the right places). In between is blank space, on a (usually) white page. And around the line drawing is more blank space, surrounded by text. Textbook diagrams show the inside of the body as being represented by acres of white space, and by a directional light – in that modern diagrams often draw only the organ in question: all else fades into whiteness. Occasionally, other colours are used as immediate background to provide contrast to a brightly coloured illustration. Thus, the three-dimensional models of the Stanford project, have a black background, contrasting with the (virtual) lighting of the models from the front.

Conceptualising the body's insides as space is, I have argued, facilitated by the practices of representation in scientific diagrams. Whether we speak of textbook diagrams showing organs or systems within a simple set of lines, or we describe the structure of the cell, *space* is what we largely see and describe. The typical cell of textbooks is a simplified diagram, an outline with a few organelles thrown in. The fine microstructure of the cell is omitted; it would simply clutter the drawing. Rather, we are presented with one each of the important internal structures – the single mitochondrion or lysosome, the short stretch of endoplasmic reticulum. All seem to float about in space, contained only by the cell membrane.[30] What consequences does this lostness in space have? What are the consequences of our seeing these abstracted, scientific, images?

One consequence is that it is very easy to think of doing a cut-and-paste with different organs. Moving an organ from a corpse onto paper, in the form of drawing, is a relatively easy transition (so too is removing the organs from plastic models). In a way, this transition begins the process of seeing organs as interchangeable, or at least moveable. But moving organs between living bodies was a dream of medicine, rather than a reality, until quite recently. Blood transfusions (a specific form of tissue transfer between bodies) have been tried, with limited success, for several centuries, and only became possible within the last century, as scientists discovered the significance of blood groups. Later, after about 1950, the first organs – kidneys – were transplanted. We are so used to seeing organs moved about on paper, in diagrams, that the first real transplants seemed to most people to be nothing short of miraculous.

Transplant surgery assumes the body to be a set of replaceable parts. If your kidneys fail, a kidney transplant may keep you going. If your heart or lungs fail, then you might be able to have a transplanted heart/lung. The limit, at least at present, is the brain. We have not (yet) come to the enormous questions that such a transplant might raise.

One factor contributing to the vision of the body as replaceable parts is, I suggest, the very predominance of the narrative of space and of the notion of organs as bounded entities within that space. The organs themselves seem, in textbook diagrams, to be atomistic, not much connected to each other, bounded. By conceptualising organs, at least some of the time, as somehow free floating and separate, it is much easier to make the move to seeing them as interchangeable parts. Each, in that space, becomes sealed off, separate from its neighbours. We can only imagine how we might have come to view the possibility of transplant surgery if we had come to focus on the interconnectedness of tissues.

Yet conceptualising organs through the representations of biomedicine may not match up to people's experiences of living the body. Indeed, we may use two sets of images simultaneously, drawing on the biomedical one in certain contexts (like talking to doctors), while using other images in other contexts. Contrasting metaphors of the heart – as pump or as seat of the emotions – provide clear examples (see Chapter 6). The doctors themselves, moreover, may present an ambivalent picture. Examining people's experiences of transplant surgery from an anthropological perspective, Lesley Sharp (1995) notes the many contradictions in the rhetoric of the medical professionals. 'Ideological disjunction', she notes, 'arises from the competing needs to personalise and to objectify bodies and organs' (p. 357).

Organ recipients must try to make sense of their own bodily transformation in the face of such contradictory messages. The donated organ has to be conceptually assimilated into the person's own sense of inner space, to become 'their' new heart or kidney. That we have come to accept these developments, alongside the rapid development of technologies involved with moving genes, is a result of the abstractions that have constructed our bodies as sets of replaceable parts.

A second consequence of the scientific abstractions is that there is simply no language within Western culture, other than science, to begin to describe our internal structures. These we know only through the historical location of the corpse, opened up to the gaze through dissection. Romanyshyn distinguishes between the place in culture of the corpse and the dead body. The corpse is the anatomised body,

> apart from the world, and a body whose interior spaces are filled with organs which define the mechanisms of life. It is a body which has developed an interior, an inside which now tells about

the outside, an interior whose space is progressively mapped and charted as locations of heart and kidneys, lungs and stomach, bones and blood, muscles and nerves. The dead body, in contrast, has no such interior space. It is, rather, a gelatinous, amorphous mass. There are no organs inside a dead body ... [Separated organs appear only] provided that we have taken up the anatomical gaze which in dis-membering the body fragments it into parts whose technical functions are disconnected from and indifferent to a living situation. (Romanyshyn 1989, pp. 128–9)

The only language we have for speaking of body parts, then, derives from that dis-membering, from the 'parts with disconnected technical functions'.

That is one reason why texts such as Monique Wittig's *The Lesbian Body* (1975) are so transgressive. It is, of course, radical precisely because it deals with the lesbian body, and with lesbian desire. But it is also radical because it explicitly eroticises the internal organs, the blood/guts/brain of the lover. And there is no easy language for this move: we are so used to narratives of inner organs as mere bits of anatomy, sitting in the space(s) of the body. Narratives of the corpse do not fit with the inner organs of the living body, especially with the body of the lover.

Seeing structures as lost in space can also be subject to a 'racialised' reading, in which the space of the body becomes constructed through whiteness (even while difference is denied by the normalising practices of biomedicine). Light directs the gaze onto the glistening tissues of the exposed inner body: dissection exposes the body 'to the full light of objective consciousness: a blinding, antiseptic whiteness; a body of purity' – to return to the opening quotation for this chapter. The direction of light – as used in photography or in portraiture – is crucial to the creation of images of racialised 'whiteness', as Richard Dyer (1997) has argued in his analysis of the construction of whiteness in film. The whiteness of the skin of the female corpse is very clearly lit in some illustrations of dissection from the late nineteenth century.[31] Her skin is pale, translucent, lit from above for the gaze of the male doctor and for the viewer of the engraving.

From dissections have come the abstractions of the diagrammatic rendering. There are obvious racial and gendered markers here – the shape of the outline of the torso, for example. These diagrams are, however, also presenting the stylised organ as outline, usually on a white page. Now it might be objected that this is merely convention, that we happen to use white paper and have black print, a heritage of printing technologies. To be sure, many diagrams and illustrations in textbooks now use colour printing (though the white background of the page remains). But what I want to emphasise is that skin is still more commonly represented by the 'flesh' colour of crayons (that is, pink) than it is by other colours, and that how we 'read' the whiteness of the page does become significant when what is represented on that page is an abstracted form *of the human body*. It is at that moment that we read the whiteness or pale colour of the diagrammatic image.

Diagrammatic images still tend to use black lines on a white background, with the exception of X-rays and the images produced by the ultrasound scan. Here, we see white lines on a black background. But in the case of the ultrasound image at least, the image is framed as a screen, not a body; moreover, its hazy mishmash of white lines must be interpreted (as women must do if they want to 'see their baby' in ultrasound images of their pregnancies). It is also usually moving about, in ways that do not necessarily correspond to what we may feel about our insides; it becomes a strange form of entertainment, in black and white.[32] Such images do not easily represent the body.

Reading the whiteness of many diagrammatic images reflects back onto the spaces in which the inside of the body is made outside – the hospital operating theatres and the physiology labs, and their spatial organisation. In hospitals particularly, whiteness can seem to pre-dominate: hospital wards may seem to be mostly white – walls, sheets, nursing uniforms – which, in the experience of one patient, felt at odds with her black identity (Rivera Fuentes 1997). White is the colour (or its lack) of hospital spaces (and thence represents the disempowerment of the patient in the hospital bed, especially if she is not white), just as it is of the spaces by which we have come to imagine bodily organs through their constructions in scientific illustration.

We can only speculate on how these images and imaginings of internal space and light affect our experiencing of our bodies. To 'know one's body' as a technical function, or as a structured space, requires distancing oneself from it. The scientific narratives do not belong to our lived bodies. Experiencing a beautiful sunrise cannot be explained in terms of electromagnetic radiation (though the technicali-ties of our seeing might do so: Romanyshyn 1989, p. 68). Indeed, it is precisely that failure of the scientific narrative to explain how we experience our internal bodies that makes the mechanical models of science so powerful: we can distance ourselves from these normalising stories. They are about 'the' body, never about *my* body. So, if I think about 'how the body works' I invoke mental images derived from my scientific training. I can easily reproduce the anatomical diagrams of the textbooks, or the machine analogies. But I do not imagine these if I think about how my body feels, about how it feels to *live* my body.

Writing about space and time in relation to the lived experience of embodiment, Elizabeth Grosz (1995) notes that, 'The subject's relation to space and time is not passive: space is not simply an empty receptacle, independent of its contents; rather, the ways in which space is perceived and represented depend upon the kinds of objects positioned "within" it, and more particularly, the kinds of relation the subject has to those objects ... Nothing about the "spatiality" of space can be theorised without using objects as its indices' (ibid., p. 92).

Now Grosz is referring here to the comportment of the subject/body *in* space. I want to read these sentences, however, in connection with the representations of bodies-in-pieces in scientific illustrations. Space – the space around the sparse lines of the diagram – is indeed not an empty receptacle. It, too, signifies. It carries connotations of race (the construction of whiteness), of gender (internal space coded as feminine by contrast to solidity), of the body as a set of parts. It connects, too, to the spaces in which we experience the body as machinic – the spaces of hospital wards for example.

I have concentrated here on the cultural significance of diagrammatic representation – the graphs and outline diagrams of the internal organs of the body – and how space is constructed by these images. In the next chapter, I turn from the structural questions of anatomy to the functional questions of physiology: how do we think about how the body works? What kinds of metaphors prevail in the scientific narratives, and how did they develop?

5

Traces of Control:
the Body as Systems

My body and I have never been friends ... I tend to think of my
body as something to anchor my head, the place where the really
important stuff is going on. Over the years, I have ... [declared]
... a kind of truce in which I agree not to subject us to further
bizarre diets and Spartan programs of torture; my body, in turn,
agreed to function without causing me severe limitations. (from
'First Stirrings', Bray 1994)

In this chapter, I want to turn to some of the predominant narratives
by which we might describe how our bodies work – their function
and physiology. The predominant metaphors of the physiological body
are often mechanistic or militaristic, as several feminist critics have
noted (for example, Haraway 1991; Martin 1994). The metaphor of the
body as factory, noted in the previous chapter, is one example, linking
the previous theme of space to that of control and mechanism.

Control: in health, most of us take for granted our bodily functions,
the way our bodies seem to manage themselves. We have learned,
moreover, to exert control *over* our bodies, trying to bend them to our
will. In Western culture, that takes many forms, such as dieting, or
exercise regimes. Rosemary Bray tells us, in her short story, 'First
Stirrings', of how her body 'agreed to function' so long as her mind
agreed not to subject it to the tortures of dieting.

Yet bodies can also be, or go, out of our willed control. We have only
to think of the sick body; in illness, our bodies seem to betray us, as the
self-maintenance that we normally take for granted appears to break
down. Lack of control also has social meaning, carrying connotations

85

of social class; controlling the body, for working class women, might for example be perceived as necessary to becoming middle class. Thus a fat body is a body symbolising the failure of control both bodily and socially (Skeggs 1997; p. 83). The maintenance of control, moreover, depends on other conditions, such as adequate nutrition; in that sense, willed control over our bodies and their appearances is a luxury of affluent societies. Whatever concepts of control mean in the language of physiological systems, they can mean something very different in the lives of people unable to control what happens to their bodies.

Here, I want to look at the development of some critical ideas in biomedicine informing 'how bodies work', their physiologies; many of these ideas were developed out of engagement of their champions with military questions in wartime, as well as out of prevailing discourses of social control. The subsequent language of control systems became a familiar narrative, both in popular texts describing bodily functions, and in scientific accounts. Control systems certainly featured strongly in my own undergraduate training – servomechanisms, feedback loops, regulation – these were the predominant motifs of my physiology education.

To be sure, I learned this language: yet I want also to reflect here on the effect it had on me, in the process of learning to become a scientist. So, in a few places in the chapter, I pause briefly to reflect on my own experiences, borne out of the requirements of a scientific training. I still find it difficult to integrate the language of science with personal reflection. But I find it equally difficult to write about those concepts of control and regulation without engaging with them as a feminist critic, occasionally pausing to think about what learning them did to me. In some ways, these parenthetic comments stand as a form of resistance to the theme of control, a way of expressing my unease.

My exploration of these narratives focuses on two main themes: the key theme is that of control/regulation, concepts deeply embedded in prevailing understanding of physiological systems. A second theme is traces/information. 'Information' is arguably *the* predominant motif in modern biology (Keller 1992, 1995; Haraway. 1997); such 'information' flows from the genetic code, or washes through the operating nervous system, leaving traces. For the nervous system, these traces are electrical. These two themes have somewhat different implications for how we view the body and its boundaries, as I will argue. As before, how we might read scientific diagrams is part of my concern here.

I do not pretend to cover all, or even a representative sample of, the principles of physiology. Rather, I focus on what I consider to be *predominant* themes, and how they developed out of particular social and cultural changes. These are the themes likely to influence how most people think about bodily function. But if so, what are the implications for feminist understandings of the body? And what are the implications for clinical practice – which is, after all, the context in which most of us will be confronted with the need to think about how our bodies work? And what about wider politics, the political consequences of particular understandings of bodily function?

The use of mechanistic language inevitably pervades physiology, as it does so many areas of science. One of my favourite textbooks when I first began teaching was a little book called *How Animals Work* (Schmidt-Nielsen 1972). I had two reasons for particularly liking this book; the first was that it was well-written, at times dryly humorous. Second, Schmidt-Nielsen's starting point was his fascination, which I share, with the adaptations of animals to their very different environments. So, he begins by noting his surprise at the number of small rodents he found on a trip to the Arizona desert. How on earth, he wondered, can such small animals survive in such an arid area, where water is very scarce? How, indeed: humans could not. Thus he studied the water balance of these creatures, how they could manage to obtain water from food, and how they conserve water within their bodies.

Yet very few animals are actually illustrated in the book, which relies on diagrams and graphs. If you want to know about what a desert kangaroo rat actually looks like, you will have to turn to natural history. The diagrams in this book, as usual, symbolise the physiology mathematically, requiring the reader to know *how* to read the illustrations.[33] Here, we find a variety of diagrams to show how, for example, the desert rodents manage to conserve water, by lowering the temperature of exhaled air (which reduces its humidity). Thus, the animal's nose becomes a diagrammatic box, with arrows to show how air moves. The text then goes on to describe an engine using heat exchange in ways similar to the nose of the kangaroo rat.

The juxtaposition of the heat exchange engine and the description of the noses of desert animals forces us to see the latter in mechanistic terms. So, too, does the widespread use of graphs and particular styles of diagrams. For example, like many other physiology texts, *How Animals Work* uses many box-like diagrams to simplify specific ideas. One such is a diagram consisting of a series of boxes to symbolise the complex respiratory system of birds (birds do not inhale and exhale as mammals do, but have a more complex airflow through lungs and airsacs). The boxes reminded me of the line diagrams of pistons used in teaching elementary physics.

These are highly abstract images, far removed from the actual appearance of avian insides. What I want to emphasise here, however, is precisely their mechanical nature. These *are* images of mechanical structures, transposed to stand as symbols of complex physiological processes. Is this any longer metaphor? Or is the reader expected to see the living body *as* mechanical structures? And what is the impact *on* the reader, as she learns to 'see' animals' insides in terms of engines? Are other readings of these texts possible, as feminists must learn to 'read against the grain'? Such questions, of course, are heresy in science; they are never asked.

One important theme of elementary physiology texts is that the body consists of a number of separate, though interlocking, systems. Again, the idea relies on mechanical imagery, implying similarity to engineering systems. We read of the circulatory system, the nervous system, the endocrine system, for instance. Usually, each of these has a component that seems to have command over others – the heart, the brain, the pituitary gland – yet can be affected, too, by influences outside of that system. Thus, the rhythms of the heart can be altered by the secretions of the endocrine system (accelerating in response to adrenaline, for example), while the brain inevitably responds to stimuli outside of the body.

A 'system', in this classification, seems to imply that there is a specific purpose served. The 'circulatory system', for example, consists of the various organs that serve to circulate the fluids of the body, particularly the blood, while the 'digestive system' digests food. But they are not always so purposive. The 'endocrine system', for instance, is not linked by function. On the contrary, it is a loose affiliation of glands which secrete a variety of different hormones into the blood. Precisely because of the language of 'systems', most textbooks describe it *as* a set of such glands and hormones. Yet each hormone has different effects within the body; several alter blood sugar levels, yet come from quite different glands. 'Systems', then, is a loosely defined term in relation to physiology: as we shall see, its persistence derives from its extrapolation from systems theory in engineering, to events within the body.

An organising principle of physiology is homeostasis, the maintenance of constancy or equilibrium. In general, the body's systems maintain themselves in relatively constant states; the internal temperature is normally regulated to stay at around 37 degrees Celsius, while blood sugar levels are kept within certain limits. It is this self-regulation, of course, on which we all rely, which enables me to ignore my bodily needs for oxygen, for example, while I sit at the computer. We generally only become aware of those self-regulated systems that require active intervention by the conscious brain; hunger pangs, for instance, might be physiological consequences of regulation by various systems, such as the ways in which blood sugar is regulated. But my response to them is conscious and cultural; I must actively do something to alleviate them.

Once food has left my plate and entered my body, however, it disappears into a void. I know nothing more about it, until whatever is left behind from digestion makes its presence felt in my gut. My body, apparently without my willing it, carries on these hidden exchanges, managing a constant balancing act.

The ways in which physiological processes are described draw heavily on systems theory, emphasising the role of feedback to maintain homeostasis or to influence particular states. Negative feedback, for example, refers to the way in which the output of a system is fed back into the system such that the output is subsequently reduced. Our bodies regulate their own temperatures, for example, generating heat by shivering or sweating to cool down. Or consider the control of parts of the endocrine system; the thyroid gland, for instance, secretes a hormone (thyroxine) that affects subsequent output of thyroxine, so that output is relatively stable. It does this by means of its effects on the pituitary, which in turn affects the thyroid. This control is moment-to-moment; as soon as the thyroid responds and reduces thyroxine secretion, the pituitary will detect the falling levels and so increase its stimulation of the thyroid – and so on.

Feedback need not, however, always be negative; positive feedback also occurs, in which the output becomes augmented. One example is the surge of oestrogens from the ovary at the time of ovulation. As levels of oestrogens go up, so they stimulate the pituitary to produce more of the hormone that in turn stimulates the ovary to secrete oestrogens. Eventually, however, negative feedback returns and levels fall. These feedback processes of systems theory have been directly exploited in medicine. Think, for example, of the millions of women who (for all its problems) take the contraceptive Pill; the Pill works by suppressing the feedback between ovaries and pituitary.

Control and regulation are key concepts in this physiology; stability and constancy are what result. One consequence of this conceptualisation is that lack of constancy is taken to indicate a failure or breakdown of the system. Fluctuations thus become constructed in these stories as verging on the pathological – which, as women have often recognised, is a problem, given the normal variations of the menstrual cycle or fluctuations with age as we reach the menopause. Indeed, this is one reason for the increasing use of Hormone Replacement Therapy for the menopause. While this may be useful for some women, some of the time, in alleviating distressing problems, its use is predicated on the assumption that the changes of menopause are in some sense pathological or undesirable.[34]

The language of feedback and systems derives principally from cybernetics – the 'science of control and communication'. The word cybernetics came to prominence with the 1948 publication of Norbert Wiener's book of the same name (and tellingly subtitled 'Control and Communication in the Animal and the Machine'). Wiener was a mathematician working at the Massachusetts Institute of Technology, and was also interested in physiological systems. In his book, he drew explicit parallels between the historical development of particular kinds of machines and the development of particular ways of thinking about the workings of the body.

From earlier analogies with clockwork mechanisms came the nineteenth century view of the body as a heat engine, akin to the great steam engines of the Industrial Revolution. But by the middle of the twentieth century, Wiener argues, the age of communication had taken over from the steam engine; electronic systems, and communication and control, now dominate. In the next two sections, I will outline some of the background to these ideas; to understand the implications of these ideas for our perceptions of our biological bodies requires that we know something about their cultural development.

Carefully regulated control, systems of the body stabilised by feedback, diagrams of boxes and pistons, a language of machine logic: cybernetics quickly took over the teaching of physiology in mid-century. What did it mean to me to learn that this was how 'the body' works? One response was fascination, the thought that bodies were so well adapted to carry out their purposes – the beauty of design. Another, however, was to separate my scientific knowledge of 'the' body from *my* body – a response to the controlling power of discourses of control/ regulation. I can speak the language of biomedicine, but only by separating it from my feelings. How my body feels to me always seems to escape the net of technoscientific language, whether in pleasure or in pain. To become a scientist – whatever the pleasures and fascinations – thus involved me in a deep fracture of my self.

REGULATION AND CONTROL: THE LIVING MACHINE

Although current models of cybernetic control are usually traced back to the 1940s, many of the key components were in place long before. Bodies had to be understood as living machines, requiring internal regulation, before they could be understood in terms of homeostatic systems. Machine and factory metaphors, not surprisingly, thus emerged as significant narratives for describing how the body works and how its inner spaces were functionally organised.

Measurement and machines for diagnosis were as integral to these shifts as the emergence of the Industrial Revolution itself. As technologies for measuring temperature improved, for example, so it became possible for doctors to use changes in temperature as an indicator of change within the body (Reiser 1978, p. 91). A change in temperature was increasingly understood as a sign of illness; underlying that notion is homeostasis, the maintenance of body states.

Diagnostic machines, moreover, changed the relationship between doctor and patient. Machines that converted what was subjectively felt into numbers and graphs meant that several observers (doctors) could discuss the output: 'Number, graph, and lesion – each connected subjective impressions that physicians personally gathered at the bedside to an objective form that permitted collegial debate and ready comparison with past, present, and future medical data' (Reiser ibid., p. 91). The experiences of the patient, which had been solicited by the doctors of earlier epochs, were no longer particularly relevant. What matters with the rise of scientific medicine are the 'objective' data of the graph.

Monitoring of bodily change relied on concepts of organic organisation. A growing concern with principles of organisation can be traced through both the natural history and the physiology of the early nineteenth century; scientific work sought to trace common principles of organisation across different body forms (through comparative anatomy; see Figlio 1976). The theme that emerges, suggests Figlio, 'is an essential inwardness or interiority characteristic of living beings; an inwardness whose traces may be glimpsed on the surface, as a physiognomy, but whose nature is always beyond the reach of visibility or total comprehension' (ibid., p. 38).

Organisation was debated in terms of design, the design of divine wisdom – a wisdom that created parallels between the interior of the body and the body politic. That is, organisation 'referred to the active inter-relating of constituent parts according to a wise plan' (ibid., p. 43), a concept of organisation that could be equally well applied to the social order as to bodily order. Change, in this organisational metaphor, was not desirable; organisation had to be maintained in both social and bodily order, notes Figlio. There is more than a hint here of the physiological controls of homeostatic systems to come later.

Present day models of physiological systems derived from cybernetics also have roots in the earlier application of thermodynamics to living organisms. Indeed, it was just that application that contributed to the nineteenth century 'triumph of reductionism over vitalism', much applauded by some contemporaries, and which led to the establishment of the concept of regulation in physiology (Canguilhem 1988). Many physiologists of the time sought to explain how living bodies work in terms of the laws of physics and chemistry; in particular, they focused on the processes of respiration, explaining them in terms of the idea, drawn from thermodynamics, of conservation of energy. If life could be explained in such physical terms, many argued, then vitalism (the assumption that there was something about 'life' that could not readily be explained by physicochemical processes) could be defeated.

This battle over explanatory frameworks in physiology, of mechanism over vitalism, partly reflected a growing professionalisation among both physiologists and physicians in the nineteenth century. Practitioners of the physical sciences were perceived as having more status, to which physiologists aspired: mechanistic explanations thus were more desirable if physiology was to become fully 'scientific'. Seeing respiration in terms of physicochemical narratives of energy conservation set the terms for the later development of control systems. As we have seen, mechanism certainly won out.

Intrinsic to these developments was the 'black box' assumption, by which experimenters treat the body or an organ as a closed system, or 'black box'. Kremer (1990), in his study of experimental physiology, argues that 'black box' assumptions were central to experimental methods in the mid-nineteenth century. Scientists could observe what goes in, or comes out, of the box (body, or organ), from which they could infer what might be going on inside (ibid., p. 14). The system is, moreover, closed. That is, if equilibrium (and health) are to be maintained, then whatever enters the system must be matched by something leaving the system; the energy of the system is thus conserved.

Not surprisingly, such ways of thinking found analogy in the steam engines of the Industrial Revolution. The 'living machine' was the new metaphor. 'It is no coincidence', notes Kremer, 'that during the decades of the 1830–40s railroads began to cross Europe, the principle of energy conservation was formulated, and physiologists began to write of the human as steam engine' (ibid., p. 18). By the early twentieth century, he suggests, the metaphor had developed further, to conceive of the human body not as a single machine but as a hierarchy of smaller ones, to form a whole factory (p. 19).

By this time, too, the use of electricity was becoming widespread, adding further to the metaphor. The relative inefficiency of the body–machine, suggested one commentator, could be improved by electricity:

> The digestive organs and the organs of assimilation, which are the boiler plant, have an efficiency much lower than a steam boiler plant. The fact is that the old machine consumes too much vital energy in preparing the fuel (food) for assimilation. Nearly all of this work can be done by artificial means, and as this is a chemical process and as electricity promotes and hastens chemical activity, no doubt this agency will be utilised as an assistant to the natural forces of the body. (Foree Bain 1910; cited in Marvin 1988, p. 142)

Factories, efficiencies, utilising electricity – the metaphor of the body as industrial production, as subject to controls and regulations is here made clear. So, too, is the prophecy of the cyborg as a product of a marriage between human body and electronic circuits – an image familiar to us today.

'Black boxes' and electronic circuitry as metaphors have remained central to thinking about systems in physiology; scientists may measure outputs and alter inputs, but the core of the system must be assumed. One physiology textbook (Basar 1976) notes, for example, that it will 'consider the biological system under study as black boxes' (p. 9), noting later that the expression 'usually refers to an apparatus, for example an electronic network, having one input and one output and performing a defined operation ... In a black box only the input and output functions are known and not the structure or processes performing the input–output operation' (p. 14). The problem for physiologists, however, is that animal bodies are not like electronic networks: they are messier. 'The biological preparations under study often have short life-times, or the parameters measured change rapidly', notes the author (p. 18)

Writing this, here, made me pause to reflect on the language and on my own responses while engaging with the text: to think in terms of 'black boxes' is to assume that it does not matter much what goes on inside the box. All that matters is the (measurable) output. The symbolic box, however, also stands for the living animal, whose potential pain is thereby ignored, figuratively written out by the abstraction. The very phrase 'preparations' glides stealthily over the processes of the experiment that ensure the death of the animal. The 'preparation' having a short 'life-time' might, for example, be a living – if anaesthetised – cat whose heart is being watched to measure the effects of particular drugs. The 'preparation' will be given an overdose of anaesthetic once its functions start to fail – at the end of the 'short life-time'. The language and assumptions cover more than they convey. I, at least, cannot ever think about 'preparations'; for me, it is always about living animals, not mysterious boxes – another disjunction between the scientist and my experiencing, feeling, self.

Yet I had to learn the language of regulation, control, even the factory metaphor, as all these ways of thinking have run through physiology for a long time. But in the second half of the twentieth century, another way of thinking about the body's workings has also become significant, sitting alongside those earlier, industrial, metaphors. Now, we can also think about the body in terms of flows of information.

LIVING CONTROL SYSTEMS: MAINTAINING INFORMATION

'Information' emerged as a central organising concept in biology after the Second World War, based on the notion that information can be quantified (into the 'bits' of information now made familiar by the widespread use of computers). What also emerged was the notion that these concepts of servomechanisms – of self-maintaining systems – could be applied to biological processes. The body thus becomes a communications network. 'Information' is now the key narrative, seemingly subsuming everything; it is certainly the dominant concept in the biology of the late twentieth century, most notably in relation to the genetic code (Haraway 1991 1997; Keller 1995).

Studying physiology in the 1960s meant that I was thoroughly immersed in these servomechanisms, in concepts of information and control; they abounded in every textbook, every lecture. Bodies became, in the discourses to which I was exposed in my training, subordinate to signal processing; to study them required at least a working knowledge of electronics.

As I have noted, several of the components of systems theory in physiology were already in place by the end of the nineteenth century. But it was in the twentieth century that it developed into the form it has today. To some extent, the lineage of basic concepts of cybernetics is well known (Canguilhem 1988). I sketch that lineage here, however, in order to trace the cultural origins of key ideas and how those influence our current understandings. As we shall see, war was a core theme in the development of these ideas, contributing to the militaristic metaphor – and hence to particular ways of thinking about the body.

Central to the development of systems thinking, and to the concept of homeostasis, was the physiologist Walter Bradford Cannon. Cannon entered Harvard Medical School in the 1890s, and quickly took up physiological research. After studies of the physiology of swallowing, using X-rays, he moved on to studies of brain trauma (Benison et al. 1987).

Initially rather reluctant to become involved with research directly related to the events of the Great War, Cannon changed his mind, following the sinking of the Lusitania. In 1916, he became Chair of the special physiological committee for the National Research Council of the US. One of the major challenges facing the committee was to investigate shock – a frequent cause of death on the battlefields. These investigations led Cannon and his colleagues to Europe, in 1917, where he began to examine the blood of soldiers in shock.

What Cannon suspected was that shock is accompanied by a profound change in the pH of the blood; it becomes markedly acidic, leading to changes in its viscosity. Injecting an alkaline solution (sodium bicarbonate) into the vein of a soldier with rapidly worsening shock produced rapid recovery, he found (Benison et al. 1991). But, for a while, cause and effect remained a mystery. Was the fall in blood pressure occurring in shock related to the marked shift in pH?

At first, Cannon had believed that the fall in pH was primary. Later experiments indicated that it was not, rather, that tissue injury led to biochemical changes that induced the fall in blood pressure and the pH shift. Both the horrors of the war, and the specific studies on cause and effect in the physiology of shock, affected him. What he had observed was the balance of the body swinging rapidly out of control, a cascade of events leading ultimately to death. Small shifts in pH, in blood pressure, were all part of a picture of an organism shifting rapidly towards destabilisation.

Cannon's work later concentrated on precisely how the organism

95

maintains stability, ideas which he brought together in his book, *The Wisdom of the Body*, written for lay audiences and published in 1932. The theme of constancy itself was not new; Cannon cited the work of the nineteenth century physiologist Claude Bernard, who had emphasised the 'internal environment'. Bernard (notable for his defence of the use of living animals in experiments – as well as his cruelty in doing so) insisted that physiology could be understood in terms of physicochemical laws. Physiologists must 'analyze the organism', he averred, 'as we take apart a machine to review and study all its works' (Bernard 1865 (1973), p. 65).

Bernard believed that events in living organisms are 'absolutely determined. That is to say ... that once the conditions of a phenomenon are known and fulfilled, the phenomenon must always and necessarily be reproduced at the will of the experimenter. Negation of this proposition would be nothing less than negation of science itself' (ibid., pp. 67–8). Bernard was writing this to oppose the vitalists, who believed living organisms not to be so determined, but having some kind of inner force that was not amenable to scientific inquiry. The language is noteworthy for the way it subordinates the living animal to the control, or whim, of the experimenter, and portrays the animal as merely a machine, completely determined.

Bernard sowed the seeds of many elements of what Cannon later incorporated into his systems physiology – most notably, the concept of the internal environment, and how that is maintained with respect to the external environment. Most organisms maintain their internal environments, but in doing so their internal environment is not at equilibrium with the external one. Terrestrial vertebrates, for instance, have to cope with, and prevent, loss of fluids and salts, while many marine species have to strike a balance with an external environment containing higher levels of salts than their own internal environments. Bernard recognised these abilities, by noting that 'living machines are therefore created and constructed in such a way that, in perfecting themselves, they become freer and freer in the general cosmic environment' (ibid., p. 79).

It is that balance, and the need to maintain it against the odds, that Cannon recognised in *The Wisdom of the Body*. Stability, he knew from his observations of shock, was maintained at a cost, and could break down catastrophically (see Cannon 1932, p. 37). 'Our bodies', he wrote in the book, 'are made of extraordinarily unstable material ... When we consider the extreme instability of our bodily structure, its readiness for disturbance by the slightest application of external forces and the rapid onset of its decomposition as soon as favoring circumstances are withdrawn, its persistence through many decades seems almost miraculous' (1932, pp. 19–20). Rejecting the term 'equilibria' as having too specific a meaning in physicochemical processes, he coined the now-familiar term, homeostasis, to describe the 'coordinated physiological processes which maintain most of the steady states in the organism ... the word does not imply something set and immobile, a stagnation. It means a condition – a condition which may vary, but which is relatively constant' (ibid., p. 24).

Yet Cannon's thinking was not only a product of his experiences at the front and his reading of Bernard. It was also a product of his (left) political beliefs (Bradford Cannon 1975), and on what has been called the 'culture of crisis' of the 1920–1930s. This combination of influences led him to extend his ideas of homeostasis to models of social stability (Cross and Albury 1987). What we can learn from the body, he believed, was the need for regulation to prevent crisis and catastrophe.[35] Thus, the last chapter of *The Wisdom of the Body* is entitled, 'Relations of biological and social homeostasis' – clearly presuming the latter.

Cannon was, moreover, writing at a time when research was proceeding apace on the sex hormones and their physiological roles. Such research was set against a background of ideas of sexuality and its social regulation (Oudshoorn 1994). The need for control of sexuality had been an important theme of nineteenth-century medical (and other) texts, and reinforced beliefs in racial and sexual superiority. Sander Gilman (1992) comments for example that 'The colonial mentality which sees "natives" as needing control is easily transferred to "woman" ... This need for control was a projection of inner fears; thus, its articulation in visual images was in terms which described the polar opposite of the European male' (ibid., pp. 194–5; see also Anderson 1992[36]). Sex and control belonged together in these discourses.

That the action of the sex hormones might be described as part of feedback controls seems almost inevitable; concepts of social control were built into the discourses of sex and the body and found their way from there into narratives of the physiology of reproduction. In turn, the concept of feedback controls in the body were founded on a heritage of racist and sexist beliefs about social control *of* the body.

Regulation and control have remained key themes in physiology (albeit now largely separated – at least explicitly – from notions of social control). One of the books I still possess from my undergraduate days is *Living Control Systems*, by Leonard Bayliss (1966). It describes 'servosystems' in physiology, and processes of 'automatic control', then leads on to how such systems respond to changes of input, and the properties of various systems of control. Control and regulation are much in evidence, key concepts of a cybernetic approach to physiology.

Bayliss' father, also a physiologist, had worked with Cannon on the problem of shock. The younger Bayliss, too, learned much from war – in his case, the 1939–1945 war, in which he had been concerned with the control of anti-aircraft guns. 'As he saw these systems develop rapidly during the war, he reflected on their many counterparts in living systems', notes the book's introduction.

Control begins the book: 'Animals and plants are chemical factories', Bayliss states in the first sentence. From there, he goes on to move effortlessly around the factory metaphor. In writing the book, aimed at an undergraduate audience, he draws parallels with industrial production and the effects of automation:

> Animals, in addition, are provided with engines which enable them to move about: the factory can move, when necessary, to its source of raw material. Just as in an industrial concern, the conversion and fabrication processes are managed and controlled ... In industry this management and control is ordinarily done by skilled men and women in the office staff and at the machines in the shops; 'automation', however, seeks to replace these men and women by 'automatic control systems'. Similarly, in the world of living things animals and plants grow and develop in such a way that each kind attains, and maintains, its proper size and shape. (Bayliss 1966, p. 1)

The 'need to control' is made clear in this extract, whether that be in a factory or in the animal body. What is glossed over, however, is that control in the factory is deliberate and purposeful; it involves hierarchies and active management. 'Control' in the living organism is not 'managed' in such ways, but reflects an order emerging *out of* particular processes, rather than being imposed upon them – a theme to which I return in Chapter 7.

Re-reading Bayliss' text, many years later, feels violating. What had begun in my life as a respect for living organisms and awe and wonder at their existences, had, by the time I got through my studies, been waylaid by such tales of living robots. Constructing living organisms as machinic bundles of servomechanisms seems to me to violate them. These narratives are undoubtedly useful as a means of telling a particular story about how the body works; as I have said, much clinical practice relies on understandings based on servomechanisms.

And the stories have their own aesthetic. But there is much more to a living organism than machine analogies. Automation hardly describes the beauty of living organisms. By re/membering living creatures, as free and beautiful in their own right, I managed eventually to regain my respect.

The machine metaphor is powerful, however. The 'factory' metaphor persisted for many decades; indeed, it can still be found, with its implications of hierarchical, managerial, control. I remember it well from many popular science magazines and books during the 1950–60s. Emily Martin describes it in her influential book, *The Woman in the Body* (1989). There, she notes that although the imagery has become diluted, several elements of it remain, even now. The cell as assembly-line factory is one example, in which DNA molecules have been said to 'direct' operations (this popular perception is, however, beginning to change, as geneticists reclaim the rest of the cell – notably the cytoplasm – as critical to understanding how development works: see Keller 1992, 1995).

Concepts of control still predominate, however, whether in terms of the information flow within the cell (from the DNA) or in terms of the functioning of the body. They are central to the idea of cybernetics (Wiener 1948). During the Second World War, Wiener worked on the development of computers and on control systems for anti-aircraft guns. 'I had become engaged', he noted, 'in the study of a mechanico-electrical system which was designed to usurp a specifically human function ... the execution of a complicated pattern of computation and ... the forecasting of the future ... [we] came to the conclusion that an extremely important factor in voluntary activity is what the control engineers term *feedback*' (ibid., p. 6, emphasis in original).

In her study of concepts of cybernetics and information in relation to molecular biology, Lily Kay (1997) stresses how many of the ideas long predated the 1940s. What happened, she argues, is that these technosciences were 'reconfigured within a new knowledge/power nexus: the military imperatives of World War II and the Cold War. In turn, these new regimes of signification served to sustain the circulation of postwar power even in fields as far removed from weapons design as heredity' (Kay 1997, p. 32) – or as far removed as the inner physiological workings of the human body.

Wiener's extrapolation of concepts of feedback to biological systems came out of his long-standing interest in Cannon's work. As a result, he began to work in neurophysiology at Harvard. Now the human body became literally incorporated into the servomechanisms Wiener had been studying in relation to guns and tracking planes. 'In defining gunnery processes as systems of communication and behavior', notes Kay, 'and a system in terms of both human and machine components, the conception of cyborg had emerged. The cybernetic organism – a heterogeneous construction, part living and part machine – germinated within the wartime academic-military matrix and matured within the national security practices of the Cold War' (ibid., p. 34; see also Haraway 1991a).

By 1966, concepts of cybernetics were spreading far and wide. A 1966 book on cybernetics included articles on cybernetic control of emotion and on cybernetics and creativity in art. The longest article extends cybernetics to 'ethical, sociological and psychological systems' (Pask 1966). Here, in research funded by the United States Air Force, we find (in amongst the mathematical abstractions) the kind of hypothetical games (zero-sum games) that later drew criticism of sociobiology. Sociobiologists were not original in their use of such games to describe evolutionary strategies, nor were feminists and other critics as far from the mark as sociobiologists objected.[37]

The 1960s also saw the neologism 'cyborg', born out of cybernetic research applied to problems of space travel and human bodies. This was to be an integration of the homeostasis of the human body and machine functions, such that bodily functions are totally automated, leaving the human mind unencumbered. Certainly, we seem to have moved closer to that dream – we now have organ transplants, microchip implants, even the possibility of attaching electrodes to nerve centres in the brain (Evans 1998). The feedback circuits of the electronic engineer are now literally moving into the physiological body.

The cyborg is, unsurprisingly, gendered. Tomas (1996) notes how the 1960s rhetoric spoke of 'man' and 'his' body – clearly gender specific. We might also add that the notion of fully automating bodily functions to leave transcendent mind free to explore itself is also deeply problematic in terms of gender, given gendered associations of mind versus body.

The figure of the cyborg has, however, been reappropriated in a feminist context by Donna Haraway, in her influential essay on that theme. Recognising that cyborgs are 'the illegitimate offspring of militarism and patriarchal capitalism' (1991a, p. 151), she argues for the cyborg – the human/animal/machine hybrid – as a potentially disruptive and hence liberatory figure. Precisely because it is a hybrid and so transgresses boundaries, she argues, it disrupts gender. Certainly, it is a potent figure, invoked frequently in newspaper stories of humans

100

with implanted microchips or other technological gadgetry. Whether or not it is a liberatory figure for feminists, however, is a matter of debate (see, for example, Morse 1994; Stabile 1994; Woodward 1994; Lykke 1996).

Whatever the potency of the cyborg for feminist interventions in science and technology, the development of cybernetics has contributed to a major shift in how we perceive the human body. Part of the reconfiguring brought about by cybernetics was what has been called the 'electronic *collectivisation* of the human body' via information technologies (Tomas 1996). Tomas argues that the concept of feedback, in particular, contributed to this reimaging; feedback was everywhere. Its ubiquity shifted attention away from notions of individualism and towards what Donna Haraway (1991a) has called the integrated circuit – flexible, network-based, boundaryless. Integrated circuits, and information flow, rely on pathways of information; within the body, a major conduit of such 'information' is the nervous system and its electrical flows.

ELECTRONIC TRACES: THE BODY'S WIRING

How did 'information' become such a key concept? What were the circumstances in which the notion of information flow became so persuasive? Much has been written about the 'information codes' of DNA; in this section, however, I want to focus on the function of the nervous system – and particularly the nerve cell – as carrier of information. I will trace some of the networks that contributed to the development of the way that we now think about 'how nerves work'. What we now associate with nerve function is usually an electronic trace, a moving line on an oscilloscope screen. Nerves work, we know, by passing electrical currents, creating a spike of voltage change, a blip on the chart.

It is these artefacts of science, the output of 'inscription devices' (Latour 1987), that help to make scientific knowledge so authoritative, as I noted in Chapter 3. We believe most strongly what scientists say when they can point to what is produced by these inscription devices. The line traced on the screen, and reproduced in countless textbooks, is a representation, standing for unseen events in the tiny chemical spaces of the nerve cell. But they are a representation that we believe constitutes proof of those unseen events.

How did such traces come to represent the function of the body's insides? What were the contexts out of which narratives of nerves as *electronic* emerge? And what are the implications of those narratives and contexts? Two interconnected conceptual developments were crucial to link electrical currents and nerves. One was the emergence of cell theory in the nineteenth century – the idea that organisms consisted of discrete units, the cells. The second was the demonstration that nerves could, indeed, pass electrical current along them.

That animal bodies could be electric has been known for centuries, particularly through those animals that can generate electric currents (electric fish, for example). But demonstrating that nerves could actually transmit the newly discovered electricity was not easy: it required the development of improved instruments with greater sensitivity.

One important factor in this history was the invention of the compound microscope, along with histological techniques for staining biological material. Towards the end of the nineteenth century, the nervous system was viewed as a syncytium, a continuous network of connecting processes (rather as we still view the histological structure of cardiac muscle). This view stemmed largely from the limits of direct observation: to the naked eye, nerve bundles appear as long, thin strands, like tiny sinews. That these, too, might consist of many, tiny, but highly elongated separate cells did not occur to contemporary observers.

(Writing that provoked further memories of my days as a biology student: I remembered again the significance of 'learning how to see', how to interpret whatever one saw down a microscope. I learned to 'see' the long thin strands as consisting of a bundle of cells. Often, what I saw down the microscope was very beautiful, whatever interpretation I learned to give it. Writing these paragraphs about the structure and function of nerves, I began to remember my undergraduate classes in neurophysiology. One feeling, vividly recalled, was my disquiet at the requirement to kill a frog (see Birke 1995). But alongside that was the feeling of awe at the beauty I perceived, even by a nature dismembered by scientific methods. This is not to condone the cruelty of the methods, simply to recognise that there is an aesthetics of nature – however culturally constructed – which I do not want to ignore, even while I learn to 'see' it in particular ways. One profoundly beautiful image I recall from my training was – is – the illustrations of developing nerve cells from studies of the embryonic brain, done by the Spanish scientist Ramon y Cajal at the end of the nineteenth century. I do not want to forget my sense of wonder at those illustrations, even in my most critical moments. Underlying my critical stance of the feminist re/analysing biology remains a deep love of the processes of 'life'. There is a profound beauty in living things, and in how they 'work', how they are put together, which always seems to drop out of the critical stance.)

Cajal's work contributed to an important shift in perception of the structure of nerve connections – and the nature of nerves as cells. Particularly important were his detailed microscopic studies of the fine structure of the cerebellar cortex (the part of the brain behind and below the main body of the brain, and which is involved in movement control). Using a new technique for staining tissues, he traced the fine branching of the nerve cell, the tiny processes that are now called dendrites because they seem to resemble miniature trees, and the long single branch of the cell, the axon.

What he also claimed, however, was that the dendrites of one cell come up close to the axon of a neighbouring cell but do not fuse with it (as would be expected of the syncytium idea). Connections between nerves – what Cajal called 'protoplasmic kisses', and today are known less poetically as synapses – are specific, not a continuous network (Taylor 1963). The network hypothesis continued to be defended, however, and Cajal's idea of 'protoplasmic kisses' was contested for some years after his work (Billings 1971). But the concept of synapses – clearly defined junctions between specific nerve cells – eventually became accepted.

Improvements in microscopy and histological staining techniques were critical to Cajal's work. But demonstrating the electrical currents of nerve cells was a more difficult task. That muscles will twitch when their supplying nerves are electrically stimulated had been demonstrated in the middle of the nineteenth century, leading to Helmholtz' suggestion in 1863 (based on experiments with such twitch responses) that the nervous system was similar to the newly invented electric telegraph. Thus the conceptual link between nerves and technological systems of communication was born.[38]

Even so, demonstrating that nerves *can* conduct electricity is one thing; it is quite another to show that they do so routinely. For some years, physiologists remained frustrated by their inability to record the secret life of nerves. Successes followed the development of new forms of instrumentation, directly out of the military research of the First World War. Neurophysiology did not look back: as a science, it blossomed, and its flowering continued to be deeply entwined with the social and political developments of the twentieth century. War has quite literally been instrumental in the fortunes of neurophysiological research – and hence in the way we now conceptualise the nervous system. (This was exemplified in my own life, even though I was born after the Second World War. My interest in electronics (on which my fascination for neurophysiology depended) came from my father. Together, we built radio sets, and I learned the principles of electronic circuitry. He, in turn, had learned his physics in the context of army communications in the Second World War.)

A turning point was the emergence of new tracing technologies, developed in other contexts. One of these was photography; another was the oscilloscope. Work intensified during the First World War on the radio compass, based on developing methods of amplifying electronic waves. This contributed to the development of the cathode-ray oscilloscope, which in turn revolutionised the study of how nerves work. Now, at last, it was possible to track fine changes in the electrical currents of nerves, and to preserve the traces as photographic images.

The image of what is now known as the 'action potential' was thus born. This appears as a short 'spike' representing a rapid change in electrical potential across the nerve cell's membrane.[39] The shift is just as rapidly returned to normal, after a brief 'rebound' or after-potential. The whole process takes about 1/60 of a second. The conceptual shift here is significant. The action potential so described is an all-or-nothing event: it is normally neither gradual nor graded. Communication between nerve cells is thus coded in terms of the frequency of bursts of such action potentials, forming the basis of 'information' in the nervous system. I will return to this point, and its cultural importance, a little later.

If military developments were important for the creation of particular instruments, military metaphors were also (not surprisingly) part of the new language of the nerve. Thus, one research paper written in 1915 noted the way that different nerve cells in a nerve trunk would be slightly out of phase with each other; so that 'rather than firing simultaneously "in a volley", [they] were discharged slightly out of phase, in "platoon fire"' (Forbes and Gregg, cited in Frank 1994, p. 220). The heritage of war is etched deep into narratives of nervous system communication.

Over the early decades of the twentieth century, neurophysiologists turned in increasing numbers to the problem of charting the action potentials of nerves. Once armed (by the 1920s) with the oscilloscope, it became possible for them to map the electrical traces of nerves. But not without difficulty. Early equipment was insensitive and dangerous; it required a great deal of expertise in engineering and electronics to work it at all, and several of the neurophysiologists of the time had had training in engineering, or had contacts with the rapidly growing electrical industry (Marshall 1983; Frank 1994).

Eventually, they did obtain relatively stable images of the action potential, which could be photographed. Analysing the developments in neurophysiology of the time, Robert Frank notes that 'What had started as a principle more like a scholastic proposition, ended as a set of pictures. Logic yielded to images created by instruments' (1994, p. 233). The concepts of neurobiology, he argues, were driven by instruments and their limitations. And what they became was 'pictures', traces across a screen or page.

Increasingly, moreover, these were instruments in sequence, in chains of other instruments. 'By the early 1950s', Frank notes, scientists were using 'four or five elements in sequence: intracellular microelectrode plus voltage clamp feedback circuits plus cathode-ray oscilloscope. The investigator could skirt total insanity only because as the number and variety of devices in the sequence multiplied, some of them became standardised and perhaps even commercially available' (ibid., p. 234).

The story of these developments in neurophysiology illustrates the importance of systems – not in the bodily sense, but in the sense used by scientists speaking of 'systems' comprising instrumentation, technique, human expertise, and particular biological materials. The narrative of the action potential developed partly out of the kind of instruments used, and the ways in which these were put together (Frank 1994). But it also developed out of the specific expertise of groups of people, that had itself been acquired through the necessities of war.

The fourth component of such systems is, of course, the biological material. Much of the work on the electrical changes of the action potential relied on the giant squid, *Loligo*. The advantage of this animal to the neurophysiologist is that, unlike many other species used in laboratories, among its many nerve fibres is a giant axon – some fifty times the diameter of most nerve fibres. The size of this axon permitted easier access to its inside surface, thus enabling scientists to measure voltage across the cell membrane. The giant squid was thus crucially an actor in the story of nerve cells and their electrical excitability.

The giant axon, coupled with the use of new techniques (called voltage clamping), enabled Alan Hodgkin and Andrew Huxley, working in the 1940s, to measure in detail the current flows occurring during an action potential (work for which they later won the Nobel prize). Apart from the significance of their work for neurobiology, it is noteworthy for the way in which it uses particular forms of visual and diagrammatic representation. Electronics – and the giant squid – provided the material.

In her analysis of this period of neurophysiology research, Maria Trumpler (1997) examines the ways in which such images became consolidated in the literature. The representations Trumpler examines are those signifying the electrical changes of nerves when active, during the action potential. What happens in the nerve axon when these spikes pass along? What makes a nerve cell electrically unstable, to make it generate a spike? One of the pivotal early studies was by Hodgkin and Huxley (1952); some components, the scientists found, were due to sodium ions, others to potassium or chloride ions moving across the cell membrane.

These movements cannot, of course, be seen directly – we can never actually see a sodium ion – but can only be inferred from what the scientists saw on the oscilloscope screen, a readout from the nerve cells connected to the oscilloscope. But the scientists also supported their argument by adding further representation; the graphical data were made more persuasive by representing the nerve membrane as an 'equivalent circuit'. That is, the researchers represented the cell membrane – a complex, biological structure whose fine structural details were at the time disputed – by the lines and squiggles that form part of the symbolic language of electronics. The symbols, of conductances and resistances, are familiar to anyone who has worked with electronic equipment. And anyone working in electrophysiology at that time had to build their own electronic equipment, as a rite of passage (Hodgkin 1992). Alongside the rite of passage in neurophysiology of pithing (i.e. killing) the frog (see Halpin 1989; Birke 1994), assembling components and knowing how they fit together in sequences of instruments was still an entry point into 'doing neurophysiology' when I was an undergraduate in the late 1960s.

While some of that imagery of electrical circuits persists, a new form of representation emerged after 1982, following new experiments with the transfer of sodium across cell membranes. Now, the images became increasingly dependent upon representations of molecules, as neuroscientists turned their attention to the molecules in the membrane which acted as 'gate' to the sodium ions. More recently, computerised models have generated moving images of gates through membranes.[40]

Trumpler documents how the earlier representations of nerve function became gradually transformed during the 1980s, as new representations emerged. At each step, scientists were at pains to show how the new representation could include the older one. If they did not, she suggests, the new model would not easily persuade others of its truth; rather, it had to be built upon the older, more accepted, model. Moreover, once DNA sequences were identified, the images derived from molecular biology began to converge with the images of electronic circuitry more familiar to the neurophysiologist.

There are several themes that I want to emphasise in the story I have just told. The first is the context in which the knowledge is generated. To a large extent, the ways in which we have come to understand how nerves work rely on techniques and expertise honed in wartime. The research made new steps when equipment designed for military use was transferred to the civilian laboratory.

Second, the sequences of instruments, forming chains, comprise a 'system' of apparatus–technique–biological material. We have come to understand physiological systems through this process of embedding; we understand 'systems' in terms of readouts from 'inscription devices', which themselves embody previous technique and expertise, and come to stand for the functioning of the biological material (or animal body) itself. As such, the animal body itself embodies layers of social relationships and cultural practices. Third, the images employed to communicate how nerves work to others, in diagrammatic form, have depended upon certain kinds of technical expertise. Specifically, generations of biology students have come to imagine the nerve cell membrane through analogies with electronic circuits and information flows.

These observations indicate that the systems of physiology are themselves deeply embedded and implicated in wider systems. The animals used, the experiments undertaken, the technologies employed, all become linked together not only in the sociotechnical systems I have noted above, but are also implicitly linked as systems within physiological and evolutionary theories. The giant squid and its evolutionary relationships are defined by scientific classifications; yet they are also defined by its location within a nexus of social relationships and technological structures. The squid's giant axon is impaled in experimental protocols on an electrode connected to a series of complex electronic devices, which further positions this species within biological categories. Systems of control operate at many levels.

SYSTEMS, INFORMATION AND CONSEQUENCES

The narrative of regulated systems, subject to tight controls, persists, alongside that of endless flows of information. These narratives, as we have seen, are deeply rooted in histories of war and conquest which inevitably structure how we have come to think about our bodies. In that sense, both narratives have histories that are deeply gendered and which pathologise difference.

The mechanistic visions of physiology rely, as I have emphasised, on concepts of self-regulating systems. Built into these is an assumption of normality, of the steady state towards which a system works. Contrasted to the 'normal' physiological body, textbooks speak of pathologies, situations in which the usual physiological controls have failed. These might include ways in which the feedback controls between the pituitary gland and, say, the thyroid become over- or underactive, with effects on the output of the thyroid itself.

Normalisation is built into the way that physiological systems have been conceptualised and described,[41] through the emphases on regulation and control. While normalisation tends to erode difference, it also serves to pathologise it. So, the focus of medicine is the *normally functioning* system/body (which itself underlies social theory). But this is described and defined *by* a physiology that itself relies on notions of normality. If deviation from the norm cannot be corrected by bodily servomechanisms, then that body must become defined as a deviant body. Thus, learning animal physiology as an undergraduate also meant knowing about pathologies – what can go wrong if the feedback mechanisms fail (in, say, an over- or under-functioning thyroid). Unlike 'normal' physiology, these pathologies were often illustrated, with line drawings or photographs. Here are images of people (occasionally of other animals, such as dog breeds with specific hormonal anomalies) whose bodies have lost their normal controls through disease.

Other markers of difference disappear, however, in the narratives of physiological control systems. We are all alike here. A heart can thus be used for transplant, whether it comes from woman or man, black or white. The normalised system is also an averaged system, minimising differences. Thus, where feedback controls shift and change with age, then it becomes a 'failure' of the system. Emily Martin (1987) notes how prevalent such language is in describing the physiology of the menopause. The feedback between pituitary and ovary is said to 'break down', leading to 'failure' of menstruation. It is not usually described as having shifted to a new set point (i.e. the point at which the system stabilises – which would be a more accurate description of the meno-pause, as feedback between ovary and pituitary shifts to a lower level). Averaging thus obscures any changes with age, and sets such changes up as pathologies, particularly in the West (for contrast with Japanese women's experiences of the menopause, see Lock 1993).

The normal system universalises, so denying different experiences and differences in physiology. Again, it is the male body that is assumed to be the standard, not only in textbook diagrams, but also in the ways that bodily exposure to chemical hazards, for example, are assessed. Risk levels rely on an assumed male worker's body (Messing and Mergler 1995). The normal system is, moreover, a concept which itself disciplines, in the sense noted by Foucault (1979), as we seek to normalise and control our bodily selves and experience illness as a loss of control over our bodies.

Yet at the same time as difference in general disappears through normalisation, so difference (as 'other' to the norm) becomes patho-logised. In this story, the development of the pathologised body becomes a tale of failure. To take one example, 'the' lesbian body has at various times been subjected to medical cataloguing and description (Terry 1995), to identify ways in which it can be distinguished from an (assumed) heterosexual version. Among these narratives are those that pathologise the physiological systems (and, more recently, genetics) of those who are not heterosexual; one major theme of such narratives is that becoming not-heterosexual represents a failure of the biological processes that are supposed to make a foetus grow into a properly heterosexual female. Thus, scientific papers may assume that lesbians must have some aberration of endocrine function, some failure of normal systems of control such that our hormones render us more masculine (for example, McFadden et al. 1998). We become not only more male but also insufficient females.

A deviant body is thus one that has failed to maintain the proper systems of control. Difference is thus minimised, except through the construction of 'others' to the normalised body; women's bodies are hence other to the supposedly exemplary bodies of male physiological controls, just as 'a' lesbian body is other to an assumed heterosexual one. 'Difference from' is built into these concepts; difference per se is not. The stage is thus set for sexist and racist practices in a medicine which quite literally embodies normalisation.

Disability, too, may be constructed in terms of deviant bodies, of 'failures' of control. The 'disciplines of normality', argues Susan Wendell (1996) in her analysis of disability and the body, 'are not only enforced by others but internalised' (p. 88). The emphasis on control that features so strongly in biomedical descriptions of physiological systems accords with the widespread cultural belief that the body as a whole can be regulated – by diet, for instance. Wendell goes on to consider the Western cultural ideal of bodily control in relation to people with physical disabilities. As she points out, much of the stigma attached to people with disabilities has to do with their bodies being out of control. This is not only the control *over* the body implied in cultural analyses, but also implies the internal functioning of the body as lacking its usual controls.

The cultural imperative for control is powerful, notes Wendell: 'I have repeatedly had to give in to the sickness of my body, to surrender to deep fatigue, weakness and pain, [yet] I still resist it every time, because the need to give in is a violation of my autonomy, my ability to plan, to make commitments, to choose. It makes me feel helpless and ashamed' (ibid., p. 113). Bodily control and discipline are valued in modern Western culture. Deeply entwined with that is the story of control written deep into the way that our physiological bodies are described in the first place.

Regulation and control are powerful themes in these stories. While the theme of control operates throughout our culture, at many different levels (Foucault 1973), it is perhaps clearest when we experience disease. For now it is the body itself that threatens to be out of control. One example of this is diabetes. Being diagnosed as diabetic can lead people into the strict regimens of control of diet and management of insulin injections. Writing about his experience of such diagnosis, Matthew Davis (1998) notes his preference for saying 'I am diabetic', rather than saying that he 'has' the disease; his reason for this choice is that 'having' can easily slide into its opposite – 'into "diabetes has me", thus admitting a loss of control' (ibid., p. 69).

The discovery of insulin has allowed those with diabetes to live. But, Davis points out, this requires constant surveillance and monitoring. What should, in a healthy person, be the unseen and unremarked function of internal physiological controls, has to become an act of will. Feedback controls can be maintained only by consciously injecting insulin. Failure to achieve 'tight' control, remarks Davis, becomes 'the patient's fault, while success derives from the physician's carefully crafted treatment regime' (ibid., p. 82). The location of the boxes by which we might diagrammatically imagine the feedback circuits has been displaced; no longer in the biological body, it has shifted to the social world of patient and physician.

Here, then, is one example of ways in which the rhetoric of feedback circuits comes to construct our experiencing of our bodies. In diabetes, the control is made explicit and exteriorised. In the next chapter, I use a case study, to turn to one particular organ, the heart; again, the otherwise unremarked systems that, in health, control our blood circulation and heartbeat can go out of control. One way of knowing that is by means of the information trace – the blips on the heart monitor screen. We might then have to face the prospect of some other forms of control, even of literally exteriorising our hearts, through transplant surgery.

My focus on a specific body part allows me to draw on some of the themes identified in the last few chapters; the heart is powerfully symbolic in our culture in ways that structure our experiencing or understanding of our internal organs – and thus structure our exper-iences of disease. Such a focus also allows me to illustrate some issues raised by transplant surgery. Transplantation follows reductionist logic. On one hand, it illustrates the successes of reductionism; after all, transplantation has saved lives. On the other hand, people facing transplant surgery do not necessarily experience their bodies in the reductionist ways that clinicians describe them. The cultural ambiguity in how we perceive the heart plays out in how we can experience heart disease or surgical intervention.

6

The Heart – a Broken Metaphor?

What the eye doesn't see, the heart doesn't grieve. (Christian Barnard, the first surgeon to carry out a heart transplant, speaking of his extramarital affairs[42])

> Life goes more smoothly without a heart
> without that shiftless emblem,
> ....
> But you've shoved me this far,
> old pump, and we're hooked
> together like conspirators, which
> we are, and just as distrustful.
>   ('The Woman Makes Peace with her Faulty Heart',
>   Atwood 1992, p. 39)

In this chapter, I want to consider one particular organ – the heart. This provides a 'case study' or intermission, to illustrate some of the themes introduced in the previous two sections, and to bring out the implications in medical practice. I chose the heart because it has a rich history of metaphoric association, as Margaret Atwood's poem attests, and because it is the site of increased medical surveillance, particularly in the West. Thus, a focus on the heart can also illustrate some of the clinical consequences of particular ways of conceptualising the body.

The heart is both the source of life (we cannot live without it) and yet it is replaceable – by transplanted organs, or perhaps partially by mechanical aids such as the pacemaker. To refer back to the main themes I discussed earlier, the heart has inner space, and occupies space;

textbook diagrams often portray it as a series of boxes, lines and arrows – part of the circulatory system. The heart is also part of systems of control and regulation; it can go out of control. And it contains our fluidity, pumping the liquids of life. I could, of course, have chosen many other internal organs as specific sites for focus. But few carry as many layers of meaning, nor have such powerful associations with life and death.[43]

The heart is gendered, not least through its symbolic association with emotionality. Even in the discourses of biomedicine, its apparently neutral status as 'merely a pump' carries connotations of gender – although, as we shall see, these are sometimes ambiguous. Not only is heart disease so often portrayed as though it were a disease uniquely affecting men (and captured in advertising campaigns urging house-wives to 'look after your husband's heart' by avoiding butter), but even the representations of that hearty pump in scientific texts can be read as gendered.

The heart, moreover, is a space as well as a muscle. Like other organs, it has architecture – rooms, walls and doorways (valves). We can see the space inside the heart most clearly *as* space when we confront the abstract diagrams of textbooks; here, the heart often becomes repre-sented by a rectangle, an empty space connecting to lines (representing blood vessels) which connect to other boxes (representing, say, the lungs or the rest of the body). Sometimes we find lines representing the feedback loops of control systems – regulating blood pressure or volume, for instance.

Alongside looking at scientific representations of the circulation, are wider cultural symbolisms associated with the heart. To write about these moves me well away from the mechanistic discourses of bio-medicine. Perhaps no other organ carries with it such a wealth of meanings. Think, for instance, of what I might mean if I tell you that I once suffered a badly broken heart. I assume you won't rush me into hospital or reach for the stethoscope, but would be concerned about my emotional well-being. I can wear my heart on my sleeve; my heart may, or may not, be in writing; my heart may literally beat faster at the sight of the one to whom I have lost my heart. The heart thus carries a great deal of cultural baggage; indeed, these meanings might be said to lie at the centre of all that we are – the heart of the matter.

Yet the heart, in much biomedical discourse, has been seen as nothing more than a pump – a sophisticated one, to be sure, but just a pump. Its rhythms are themselves subject to feedback controls. If broken, the pump can sometimes be mended through bypass surgery, or transplants, just as broken machinery can be fixed. This kind of heart was once central to meanings of life and death; doctors defined death when the heart stopped beating. But even that centrality has now disappeared: in the age of transplant surgery, the brainstem is more important, and hearts may go on beating even after brain death.

This chapter, then, explores both sets of narratives, alongside each other. My purpose is to contrast them, and thence to ask how such contrasting ways of thinking the heart might have impact upon our experiencing the heart as part of our bodies. Perhaps at no time is that contrast more stark than when a person is faced with the prospect of heart surgery – and particularly the prospect of receiving a heart from another body.

The metaphor of mechanical pump is, however, no longer adequate. For now scientists are focusing increasingly on the electrical activity of the heart, as I will discuss. We might tend to think of a regular heartbeat as an indicator of good health: but now even that story is changing. Some degree of chaos in the heart's functioning is now seen to be desirable, and too much regularity warns of impending failure – the calm before the storm of cardiac fibrillation. The heart, it seems, is yet again being redefined, now in terms of its potential for electrical chaos.

METAPHORS OF THE HEART

We have become familiar with the images of heart failure through popular television series, such as *Casualty*, or *ER*. In most episodes, there is a cardiac arrest, to which staff come running; someone wields electroshock plates to kick-start the errant organ. We watch, expectant, as the heart monitor registers the irregularities. We know how to interpret the steady electronic line. Such stories remind us how used we have become to the drama of heart failure – at least as the tale unfolds through graphical display. They also dramatise the heroic stuff of modern medicine. Like Dr Frankenstein wielding electric shock, the modern hero in white coat will – we hope – save the patient. Meanwhile, the technology reads us the story of what is happening within. The heart responds to the hero's gallant efforts – or not.

As the heart becomes a pump, so medical practice, accordingly, treats body parts as simply replaceable bits of machinery. A faulty heart can be replaced. Viewing the body as machine also contributes to a growing market in body parts (Kimbrell 1993) – be they organs for transplant, blood, sperm, or even eggs or embryos. It also permits the use of organs from non-human animals, through xenotransplantation.

Crucial to medical conceptions of the heart is the normal/ pathological binary to which I referred in the last chapter; concepts of normality are built into the notion of control systems themselves. Hence, the pathological, the diseased, becomes something to be guarded against in preventive programmes, or to be excised in surgical interventions. The diseased heart is monitored, under constant surveillance for potential breakdown. Becoming ill with coronary heart disease leads to increased medical surveillance.

Surveillance requires vigilance, however. It is the heart itself which is the focus of much of the surveillance of health to which we are subject. The normal heart may be a somewhat unmarked category of biomedicine, but we lay persons are still encouraged to work at the process of preventing heart disease. In such bodily discipline, the heart is a focus of attention. We must eat the right foods; we must take exercise. Think, for example, of the growth in 'keeping fit': a key notion in 'keeping fit' is that the heart is 'toned up'. When I once asked a fellow hiker the best route up a mountain, she asked me whether I wanted 'a good cardiac workout' or a more gentle path. The cultural imperative now is to keep our hearts in good shape along with our bodies. The heart monitor becomes important equipment for that surveillance.

Yet the irrational body – the other half to the (rational) mind – includes the heart, which is at once both symbolic (and symbolic of emotionality, not reason) *and* material. What medical practice around heart disease focuses on is precisely the unpredictability (irrationality) of a heart gone crazy. In a heart attack, the electrical excitability of the heart and the contraction of its muscular walls, are both wildly out of control. The messages of both prevention ('look after your heart') and of treatment (surgery) are about control and management. And alongside the eternal vigilance to ensure our pumps are working well, there is that darker, more irrational, set of images – the heart of emotions, of love and passion. Not surprisingly, this story of the heart is rich with metaphor and allegory. There is a powerful irony about the contrast between the story of the heart as part of control systems, with their deep associations with militarism, and the history of the other heart, the heart of love.

In the cultural association of women with emotionality, the symbolic heart becomes quite clearly gendered. This is particularly clear in several illustrations from the late nineteenth century, depicting doctors dissecting corpses. A recurrent theme features a beautiful, but dead, young white woman lying draped suggestively on a mortuary table;

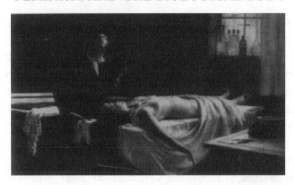

Figure 6.1 An anatomist gazing at the heart he has just excised from the corpse of a young woman. Note the use of light to emphasise the woman's body. Photogravure by R. Schuster, 1906, after a painting by D. Simonet, 1890. Courtesy of the Wellcome Institute Library, London.

she is usually bathed in light.[44] The (male) doctor has begun his dissection, and in one such picture holds aloft her heart. Her heart epitomises the seat of her femininity and beauty – even while it represents the act of anatomical dissection that lies at the heart of the mechanical model.

The heart has a long history as a signifier in, for example religious texts (Sauvy 1989). In these, images of hearts may represent salvation or temptation, to become either the 'abode of Satan or the temple of God'. While both men and women are represented in these iconic hearts, notes Sauvy, they appear in different guises: one series, from the late nineteenth century shows, for instance, male faces above the 'bad hearts', while faces of women are associated with 'attrition, contrition and perseverance' (ibid., p. 205). Racism also prevails: devils, predictably, are black, while angels – residing in the pure of heart – are feminine, long-haired and white. Faces representing the devil's temptations are rarely white.

Such images are powerful symbols, standing not only for meanings of 'the heart', but also incorporating a whole series of binaries. Among other things, the heart in these iconographies stands *both* for the interior of the person (emotions and desires contained within) *and* for the embeddedness of those representations within a wider, external, world. That is, the figures contained within the heart are simultaneously assumed to be manifest in the world outside the body – in the form of the devil, for instance. They thus serve symbolically to construct the inner space of the heart within the body as representing an outer, social, space. This ambiguity of the symbolic heart, as both

116

inside and outside the body, is a theme to which I will return below.

While reading Sauvy's text, I recalled a TV programme I watched in the USA a few years ago – one of those evangelist kinds. The preacher told how, driving along an interstate, he had experienced a great pain in the middle of his chest. He pulled over, worried about a possible heart attack. But, he then realised, 'this was Satan tempting him' from within his (literal) heart: so, he told Satan where to go, by running up and down the roadside. Satan presumably vacated his heart: he evidently did not need cardiac resuscitation before his televised appearance.

LEARNING BY HEART: GENDER AND RACE IN SCIENTIFIC METAPHORS

Thus far, I have sketched the contrasting narratives of the emotional/symbolic heart, and of the mechanical heart. Both are overlain with gender, race, and other categories of exclusion. Londa Schiebinger's analysis of gender in the history of science has shown how gender and race were read onto nature (Schiebinger 1989, 1993). From the skeleton, the discovery of other species of apes, and the classification of mammals, to the botanical details of flowers – stereotypes of gender seemed to be everywhere. The heart could not be innocent of such reading.

The metaphoric descriptions of the heart and circulation pervading scientific texts are generally mechanical or domestic (drawing parallels with domestic central heating). Generally, the heart as seat of the emotions does not appear in scientific texts – unless, that is, the texts are written for popular consumption. In that case, the miracles of modern science are prominent, but tinged with heartfelt gratitude. Here, for example, are some extracts from a popular science book, which begins: 'The heart is a life-giving pump, a simple machine with a sacred mission. Its labor is brute, its fabric coarse, yet the heart connects and sustains the body's work ... The heart is the center of life ... [which] William Harvey [who discovered that the blood circulates in the body][45] ... deemed ... 'the sovereign' of the body. The heart kindles and keeps life's flame – and so inspires man's awe ...' (Davis and Park 1984, p. 7).

This brute labourer, wearing coarse coat, brings to mind (again) the thumping machines of the Industrial Revolution; yet, it is also sacred, sovereign, and awe-inspiring. Thus, the same text elsewhere describes how blockage of the coronary arteries leads to a heart attack: 'The coronary arteries are a crown that keeps the heart alive. Blocked, they prove the king all too mortal' (ibid., p. 47). Either way, the metaphor is gendered male – a contrast to the emotional heart.

The relationship of the heart to the rest of the body is, moreover, one of 'centre' which sustains the 'body's work' in many represent-ations. Here, again, is the widespread metaphor of production: just as the pituitary was dubbed the 'master gland' of the endocrine system in many popular physiology books, so the heart becomes the powerhouse supplying the fuel to the body's periphery. The reader must locate herself, somehow, into this narrative of virile industrial production.

Let us retreat from the industrial age for a moment, and look at the early modern period – to the beginnings of what we now call the scientific method to detect the imprint of gender and race on the mechanical models of modern medicine. During the early seventeenth century, William Harvey carried out a great number of experiments on the circulation of the blood in both living and dead animals. He developed a research programme at Oxford, which flourished to become part of a new emphasis on experiment. Indeed, he could be said to have been one of the 'founding fathers' of modern science. Steven Shapin notes how Harvey thought it 'base' to learn about nature through books (as classical philosophy had done); far better, he averred, to use 'the book of Nature' (Shapin 1996, p. 68).

Reading Nature's book, however, meant interpreting the text in terms of mechanism, captured in the rhyme coined by appreciative colleagues:

> There thy Observing Eye first found the Art
> Of all the Wheels and Clock-work of the Heart:
> The mystick causes of its Dark Estate,
> What Pullies Close its Cells, and what Dilate,
> What secret Engines tune the Pulse, whose din
> By Chimes without, Strikes how things fare within.
> [cited in Davis and Park 1984]

Here, even the 'mystick causes' become reduced to the clockwork of the heart – mere mechanism that can eventually be understood by the 'observing eye' of science.

It was during this period that experimentation became consolidated as the *modus operandi* of science. Moreover, it became consolidated as a specifically *public* enterprise (Shapin and Shaffer 1985). Experiments could be seen by anyone; but not everyone could act as responsible and reliable witness – not everyone's word could be trusted, but only that of what Donna Haraway calls the 'modest witness' (1997). This figure is inevitably male and European. The modesty, suggests Haraway, 'is one of the founding virtues of what we call modernity. This is the virtue that guarantees that the modest witness is the legitimate and authorised ventriloquist for the object world, adding nothing from his mere opinions, from his biasing embodiment. And so he is endowed with the remarkable power to establish the facts' (ibid, p. 24).

So, it was Harvey's act of witness, his 'observing eye' which dispassionately 'first found' the heart's mechanism, according to contemporaries. But the early modern period was also a time of great social turmoil and passion; gender and race (among other categories) were at issue in the origins of modern science (Schiebinger 1989). The modesty, required of the witness, became increasingly linked to the (masculine) mind. It was the basis of the dispassionate stance.

I have outlined this here because I want to emphasise the significance of medical discoveries about the heart and circulation in the historical development of the scientific method. Not only the pattern of circulation, but also the first attempts at transplant were done in this period (for example, by means of attempts at blood transfusion in the late 1660s: Guerrini 1989). That we can distance ourselves to look 'objectively' at the clockwork heart, that we can accept the metaphor of mechanism, depends on that history. The emotional heart became truly separated from the mechanical heart, objectively observed.

Gender, and other categories of relation, were deeply embedded in the very scientific methods of experimentation that were at the birth of modern science. At the very least, the modest witness had to belong to certain categories of class, gender, sexuality: as Haraway wryly notes, 'God forbid the experimental way of life have queer foundations' (1997, p. 30). Among other organs, the pumping heart is not innocent of gender, nor race, nor colonialism, in its metaphorical associations.

TO MY HEART'S CONTENT: METAPHORS IN MEDICINE

The heart, then, is a powerful metaphor in many texts, carrying with it sexism, racism and the history of colonial conquest. In the narratives of biomedicine, we meet quite different and largely mechanical images. If it is 'only' a pump, moreover, there should be no particular ethical concerns about moving it from one machine to another. In this case, the machines in question are a dead human body and a living one. Such rhetoric is relatively recent. Prior to successful heart transplants, the heart 'was seen as the centre of the soul' (Adrian Kantrowitz, US heart surgeon 1996). As hearts became exchangeable, so a new definition of death developed – and the heart became consolidated as 'just a pump'. Kantrowitz described how the use of anencephalic infants[46] (born without a brain) as organ donors for transplant surgery facilitated this redefinition: 'Just because the heart [of the anencephalic infant] beats doesn't mean it's a living human being', he said (Kantrowitz 1996).

For a moment, let us consider what hearts do, in the language of the textbooks. Hearts pump. In doing so, they move blood around the body to supply every part with oxygen and nutrients. The heart of vertebrate animals develops from a convoluted muscular tube; this suffices for fishy needs, but for land animals a more powerful pump is needed to keep up the oxygen supply. Frogs, for instance, have three-chambered hearts, while mammals have four. Tiny animals have tiny hearts; but they must beat faster. A shrew's heart at rest beats ten times a second.

To achieve coordination across the heart muscle, the heart's cells are linked tightly together, and the electrical pacemaker[47] kickstarts the cycle of contraction, the familiar heartbeat – dub-dub, dub-dub. Most of the time, this goes unnoticed – it beats away at its heart's content. Sometimes, we do notice: we pay attention to it if we want to know how fit we are, or if we confront something which scares us. Then, the blood pounds in our ears.

The pacemaker acts as the initiator, the controller of the heartbeat. The heart itself is the central part of the circulatory system, whose function is to push blood around the body to supply oxygen and nutrients to various tissues. It, too, is subject to other regulations: blood pressure and volume, for instance, are usually homeostatically maintained.

To learn about the heart and circulation in textbooks, we usually come across a sketch of the heart, a shaded drawing of what a heart 'really' looks like when removed from the body. Then, there are diagrams of the circulatory system. Now the heart becomes represented as interconnecting boxes, to represent the different chambers, in turn connected by arrows indicating blood flow to other boxes – the other organs of the body. This transition – from realist drawing to diagram – involves a further abstraction. In imagining the heart as part of a system we abstract it into the spaces of the diagram. The diagrammatic heart sits in a kind of empty space; it is no longer the pulsing flesh of a beating heart.

In writing this, I have been using the language of certainty, and of causality and control – 'this is what happens' in the scientific story. It is also the language of mechanism – muscular tubes, pumps, failure to contract, cycles of contraction, pacemakers – and the language of triumphalism, in the conquest of the land by animals able to develop more efficient hearts and methods of respiration. Textbooks also reveal, relatedly, the language of medical success, of triumph over the killer coronary heart disease (CHD): bypass surgery, coronary angioplasty, transplants. We could illustrate with labelled diagrams, or histograms showing epidemiological data; the frequency of CHD by social class, for example, or of heart disease as 'responsible for nearly 50 per cent of deaths' (at least among white males in the Western world).

Scientists are, of course, trained to avoid any implication of agency in their writing – their own agency, or that of nature. But that cannot always be done; nature is sometimes recalcitrant. Here, for instance, is an example in which the heart is clearly trying to escape the strictures of scientific writing: 'When atrial contribution and the A wave kick are lost ... there is an increase in the mean left atrial and pulmonary venous pressure *in an attempt to maintain* the same level of end-diastolic pressure and cardiac output' (Parmley 1979, p. 1064; my emphasis). The heart 'tries to maintain' pressure, even though it is only a pump. It becomes an organ with a purpose.

Whatever the claims to represent reality, scientific narratives are just as replete with metaphor as the religious ones. Plumbing metaphors abound around hearts:

> despite [its] punishing workload, the heart loyally thumps three billion times in a long life, forcing blood along 60,000 miles of pipes – a plumbing system longer than the road networks of Wales and Scotland put together ... A giraffe needs a frighteningly high blood pressure to service its lofty brain – and so must construct extra strong pipes to avoid blowouts. (Young 1992)

Young gives these examples in a review of a textbook about the circulation of the blood; its author, he notes, found inspiration 'around the house' for the plumbing analogies. It is a book, moreover, which transgresses the boundaries of what can go into a science text: 'It is certainly the only physiology book I know that includes instructions for cooking its principal subject matter', notes Young (ibid.).

The plumbing metaphor reinforces the image of an organ that is quite simple, just like a pump. It is reminiscent of both industrial power (an oft-recurring image), and of simple domesticity. Even blocked arteries seem to resemble the way the kettle furs up if the water is hard. You do, of course, need the plumber's help when things go wrong, but there is a lot you can do yourself to stop the kettle furring.

Advanced textbooks are somewhat less likely to refer directly to plumbing systems. Yet they, too, rely on such metaphor. My under-graduate physiology books, for example, had long sections on the dynamics of fluid flow along blood vessels; we had to read the physics of fluid motion, to understand the principles of turbulence for example, if we were to understand how blood moved around the body. Or perhaps we could calculate the mechanical efficiency of the muscle of the heart. That is not to say that the physics is inappropriate; it clearly does help to explain how blood circulates. But, as I argued in the previous section, these frameworks of explanation draw on principles derived from engineering. The plumbing metaphor seems almost inevitable.

The parallel developments in steam powered engineering and experimental physiology in the late nineteenth century ensured, as we have seen, that scientists would perceive similar principles in each. We are now so familiar with these frameworks of explanation that it is hard to think outside them. Yet I wonder what cardiovascular physiology would have looked like if engineering had taken a different route? Perhaps we could imagine a gentle flowing stream, or invoke the fluids and flows of earlier cosmologies? We can only speculate.

And what about gender? Is the scientific heart as replete with gendered metaphors as the symbolic heart? The visual images in medical texts are rarely overtly gendered, in the way that (say) the images of religious symbolic hearts are gendered. But the written narrative of scientific accounts is different; behind the objective stance and passive voice lurks a familiar metaphor, with intimations of sexuality linked to different parts of the heart. Here, we find that the ventricles (the lower two chambers of the heart) are made of different stuff from the atria (the top two chambers). In Parmley's text, we read for example that: 'The atria are thin-walled shallow cups ... [which] serve a reservoir function. [Later in the heart's cycle of contraction] the atria serve as a conduit for the continuous flow of blood from the veins to the ventricles' (1979, p. 1063). The text goes on to tell us that the ventricles are pumps (though the left differs somewhat from the right), whose job is the ejection of blood.

The metaphors become clearer in texts written for lay readers. Again, the blood seems to flow through the atria, although these 'shallow cups' can help a bit by squeezing blood through the 'doors' of the heart valves, and into the ventricles. And now the gendered and sexualised reading of the narrative becomes clearer. The ventricles, we can learn, respond to 'the swelling surge of blood' by beginning their muscular contraction. The pressure 'forces blood upward and thrusts open the semilunar valves … The right ventricle ejects … blood into the pulmonary artery' (Davis and Park 1984, in a chapter entitled 'A Surging Pump').

This seems to be a remarkably virile ventricle, forcing and thrusting; it is always well-muscled, and reminiscent of the thrusting machines of late nineteenth-century industry. By contrast, the atria are conduits, or rooms, receiving blood, perhaps helping a bit by squeezing. To function, it seems, the heart has to be a happily heterosexual union of virile ventricle and acquiescent atrium.

Sometimes, however, even happy hearts need marriage guidance. The Davis and Park text notes that:

> The parasympathetic nervous system acts as the heart's physiological conscience. It slows the heart, counseling it to save energy. Sympathetic nerves do the opposite. They quicken the heart in answer to the challenges that come rushing on the wings of daily experience. Heat, light, love, danger – such stresses force the sympathetic nerves to step in and spend some of the energy the parasympathetics have been husbanding. (ibid, p. 65).

These examples come from books aiming to popularise the science. As such, they have power to influence how we all understand our bodies; and they draw on metaphors and analogies used in the original scientific work. Gendered metaphors abound, it seems, in all these narratives.

Like domestic plumbing and happy marriages, the beating heart can go wrong. It needs nourishing and has its own blood supply. If this gets blocked, the muscle cells can starve. The heart fails to contract properly. A heart attack follows. Because the heart is little more than a pump (it does not carry out any complex biochemical jobs, unlike the liver), it can be repaired relatively easily, so medical reasoning goes. Bypass surgery detours the damaged blood supply; coronary angioplasty enlarges the vessels; transplants may replace the whole heart. And now, with xenotransplantation (the use of organs from other species), we can use a heart from a pig. It is, we must assume, only a pump, wherever it comes from.

The heart is ambiguous. It is at once 'part of the plumbing', and something which can be tuned up. It is at once merely a pump, and a part of the complex systems and feedback controls of late-twentieth century integrated circuits (to paraphrase Haraway 1991a). It is precisely that ambiguity which allows the heart metaphorically and literally to link the apparently diverse discourses of emotionality and pumps; emotional events can, after all, alter heart rates (see Bendelow and Williams 1995). Indeed, it is just that effect that is important when, for example, animals are brought into hospital wards for therapeutic purposes: stroking the animal reduces heart rate and thus helps to reduce the risk of another heart attack.

What focuses our attention, however, is the apparently unambiguous trace of the ECG (electrocardiograph).[48] Electrical signals pass across a screen: that is the heart. The familiar sound accompanies the blips on the oscilloscope screen – or, in the television dramas, silence accompanies a flat line. And we know how to interpret it. This is another example of an inscription device, a machine for measurement in the laboratories or hospitals, whose output is the arbiter of truth (Latour 1983).

What I want to emphasise here is how we have all learned culturally (not least through TV dramas) to 'read' the ECG trace. It is not only the doctors/scientists who read it (although they must concur in their 'expert' reading): readings of the ECG are, to some extent, possible for all of us who are used to technological medicine. The machine is an inscription device, in Latour's sense, although not always located in an experimental laboratory. If the hospital doctors want to convince us that our hearts are 'normal' they have only to point to the trace (it does not matter that we are unlikely to know how to interpret all the different squiggles). They do not even have to be in the room at the time (and with the advent of new communications technology, they may even be in another part of the world).

The ECG machine, as an inscription device, does two things. First, it provides the readable format, the written trace (although we have to learn to interpret it, according to specific conventions). Second, it profoundly restructures the social relations between doctor and patient, and between patient and machine. Doctors no longer 'read' the body of the patient; instead they read the output from electronic devices.

It is here that the metaphors can come together – indeed, become thoroughly mixed up – in the troughs and peaks of the electrical output, measured by the ECG. To quote from Davis and Park's popular book again, we find a traveller's tale:

> The trace of the ... needle, its endless traverse across cardiac hill and valley, measures the beat of the resting, faithful, heart. But along this route peaks can suddenly rise, ravines can plunge. These pronounced pulses in the cardiac rhythm are landmarks of the versatile heart, the organ affected by emotion – joy, fear or rage. (Davis and Park 1984, p. 61)

From pumps to peaks to plunges to passion – a versatile organ it is indeed that can support so many metaphors.

Electrophysiologists have spent much time analysing these peaks and ravines, to derive a picture of the heart's functioning both in health and disease. Like measurements of the action potential of a nerve, careful assessment of the distances between different parts of the complex electrical output of a heart provide detailed topography. So, alongside the pump metaphor there now emerges a metaphor of the heart as a kind of power station, generating its own electrical circuits.

Reading an ECG is not an easy task; like looking down a microscope, it must be interpreted. Electrodes must be placed in particular spots on the body, as different locations yield different readings. Adequate functioning of the heart thus becomes defined by reference to the placement of electrodes on the skin – an electrical cartography of the body surface. Each electrode thus takes different 'readings' from the heart, so that, as one of my physiology textbooks put it, 'aVR [one of the electrodes, located on the right shoulder] "looks at" the cavities of the ventricles ... while aVL and aVF [other electrodes] look at the ventricles' (Ganong 1973, p. 398). 'Looking' here, of course, takes the form of picking up electrical changes, an electrical mapping of the body.

The ECG as a machine reading from the surface of the body also serves to map internal space. Its output *is* the heart. Any other electrical signals, from other parts of the body, are interpreted as 'noise', and ignored by the electronic averaging of the machine. In that sense, the body wired up to an ECG is a cyborg, part-human and part-machine; it reconstructs the inside of the body to become electrical patterns. This heart is not a pump, nor is it the seat of emotions; rather, it is a space where electrical waves collide.

So in this narrative, the heart has been defined in terms of its electrical circuits. Reading the heart then requires training, not only in reading the peaks and troughs of the ECG, but also in placing the electrodes in the correct terrain. The patient may have learned the significance of some readings of an ECG – the still line, for instance – but most of us are not likely to be able to read the detailed topography appearing on the screen. In the hospital ward, this particular inscription device serves not only to persuade those others who (believe they can) understand its traces, but also serves to marginalise the person to whom its electrodes are connected.

However efficient an electrical generator it is, the heart can be short-circuited, in cardiac arrhythmias. Here, the electrical activity is no longer so coordinated, perhaps because of reduced blood flow to part of the heart muscle; as a result, the electrical changes do not occur in rhythm and the muscle starts to contract arrhythmically. Danger threatens, and the heart begins to have a life – even a mind – of its own:

> The tissue grows anxious, 'irritable', impatient, and its cells depolarise [lose the potential difference across the cell membrane] on their own initiative ... A flurry of such beats can sometimes build into arrhythmic frenzies called flutter and fibrillation. In atrial flutter, the chambers ... maintain an orderly rhythm. Fibrillation, by contrast, is chaotic' (Davis and Park 1984, p. 103)

Chaos, flurries, frenzies – a heart beating out of control. It seems, then, counterintuitive to look to chaos theory for a way of rescuing the heart from such frenzy. Yet that is what scientists now do, in a new turn of the story of the heart as electrical. Chaos is not simply random; rather, it is persistent instability. Heart rate, moreover, can be chaotic even in a healthy heart; indeed, it is when a heart is going into fibrillation that it loses that fingerprint of chaos. Its electrical impulses become too regular and predictable.

Sometimes, the electrical wave that propagates across the heart gets 'stuck', perhaps because part of the muscle is diseased. If so, then it might begin to rotate around the heart muscle as a spiral wave, possibly then breaking up into smaller spiral waves. These waves are, researchers say, implicated in the onset of fibrillation. To deal with it, research is focusing on how small electrical signals to the heart (unlike the massive shock of existing defibrillators) might encourage the heart to sort itself out. The heart with a life of its own must be disciplined, just as a misbehaving person might. 'We are hoping this technique will use the energy of the harmful behavior to move the heart back into good behavior', said William Ditto, a researcher working on hearts in chaos, 'Rather than fight the chaotic pattern, we want to have the chaos do most of the work for us'.[49] The heart, it seems, is normally rather more chaotic than previously thought, and increased regularity may portend its breakdown. In this new narrative, moreover, the heart is becoming an organ that does, indeed, have a life of its own; it can regulate itself.

The next step in such research is from women in labour. Might there, the researchers want to know, be an indicator of forthcoming foetal distress hours, even days, in advance? And if the foetal heart starts to show too much regularity, is that a sign of 'disaster to come'? Here is yet another potential intervention aimed at monitoring foetal welfare which feminists need in turn to monitor.

In this reconstruction of narratives of the heart, we have moved from the thumping machinery of the pump to an electrical field with a mind of its own. The heart has become an actor within the body, subject to disciplinary action but equally capable of resistance. Some recent work in science studies has drawn on Actor Network Theory, which analyses the ways in which scientists enrol others – including non-human others such as animals or machines – as part of their techniques of persuasion (see, for example, Garretty (1997) in relation to heart disease). These other 'actors' form part of the networks by which a story is made.

In the case of heart monitors, it is the machine itself that constitutes one of the actors in the drama, through which the doctors maintain the power to define normal/pathological, health/sickness and so on. Indeed, in televised hospital dramas it often takes the leading part. Similarly, in the case of transgenic pigs created for transplant surgery, the pigs themselves might be seen as benevolent actors, 'helping' the surgeon – or the patient (Birke and Michael 1998).

The heart itself can be thought of as an actor. It can be enrolled – by means of keeping fit, for example – but it is also one which can resist – as shown by changes in the heartbeat for instance. The image of the heart as an active agent is, as I have just noted, becoming quite explicit in recent studies of the electrical storms of fibrillation.

So, coming to know the inside of the body (in this case, the heart) requires a complex process of establishing networks between, say, a doctor, a patient, her heart, and the machine to which the heart/person is connected. These networks become a social manifestation of the linear diagrams of control circuits, exerting their own disciplinary power. It is this complex network, not the person by herself, which 'reads' normality or pathology into the ECG trace.

That we might think about the heart as an actor was brought home to me by another webpage I found, entitled 'Women Lead with Their Hearts: A Women's Empowerment Seminar'.[50] Below the title is the ubiquitous ECG trace, followed by a subhead, 'The Science of the Heart'. The text goes on,

> Women naturally lead with their hearts. It's why women are good caregivers, managers and friends. But all too often the heart is perceived as a weakness, not as a source of wisdom and power ... Scientific discoveries over the past thirty years have revealed that the heart has a unique intelligence unto itself. The heart is central to a whole array of processes that affect brain function, personal effectiveness and overall well-being ... The Institute of HeartMath's ... pioneering *research* shows a relationship between feelings in the heart and mental/emotional balance, health and productivity. (IHM Webpage 1997; emphasis in original)

Later, we are told that we can 'Fine tune and power-up what you, as a woman, have naturally – your intuition and your heart intelligence'. Alongside the text, too, there are ECG traces. These are labelled as 'frustration' (a messy trace) and 'appreciation' (an altogether neater one). Reading the electronic traces now takes on the task of bringing together the metaphors of pumping machine and lovesick seat of the emotions. Here, then, is the heart *as* actor, and as actor which can be fine-tuned for 'peak performance', rather in the way that discourses of immunity now emphasise the need to tune up the immune system (Martin 1994). It is, moreover, an actor that is clearly gendered in this text.

Now the heart-as-pump is typically portrayed as being similar in all of us, no respecter of social categories. Even if the illustrations of body outlines in medical or biological books tend to be largely male, white and young, the internal organs bear no such obvious representation. But when we start to see the heart in medical discourses and practices as an actor, along with the appropriate machines, then it is no longer such an innocent organ.

I want now to turn from the changing metaphors by which the heart and circulation have been described in scientific narratives, to considering how those narratives might impact upon people's experiences of cardiac intervention, particularly cardiac surgery. The two central metaphors – of the heart as seat of emotions versus mechanical pump – must become separated if (masculine) mind is to gain greater ascendance. By separating mind from body, bodily invasion (in the form of surgery, for example) is made easier. The medical discourses of plumbing work to persuade us – literally – to open our hearts to the scalpel. The reductionist rhetoric reminds us that we are, after all, merely a collection of replaceable parts (a conclusion made easier by perceptions of organs in inner space, isolated from each other, noted in previous chapters). But if we understand the heart to be the centre of our emotions – the heart and soul of who we are – then we would be much more reluctant, particularly if the donor of a heart is not human. How much do we, as patients or potential patients, really believe the logic of reductionism?

OPENING OUR HEARTS: SURGICAL SUCCESSES?

Partly because of the power of the (symbolic) heart as metaphor, the discourses of biomedicine must work hard to persuade us to open our hearts to the scalpel. In a study of people undergoing coronary bypass surgery, Radley (1996) describes the joint investment, by patient and staff, in particular medical ideologies. So, success may be stressed, while problems (such as postoperative pain) are played down. Radley notes: 'In the case of coronary grafting, this [ideology] is specific in its use of "plumbing" metaphors to do with the blocking and unblocking of arteries. It also shares certain features with surgery in general in its parallel with the repairing, rebuilding or remodelling of machines' (ibid, p. 133).

Yet even while medical literature describes the body as a set of replacement parts, surgically moving organs around is troublesome for our sense of self and identity. Cultural beliefs about selfhood may conflict with notions of body parts. Indeed, there is often active resistance to reductionist rhetoric, as Emily Martin noted in her study of women and their understanding of how their bodies work (1987). Medical personnel put great stress on objectification – the heart as 'only a pump'. Yet recipients of transplanted organs tend to experience conflict between this mechanistic/reductionist view of the body and wider cultural beliefs (Joralemon 1995; Sharp 1995).

'Domino' procedures (in which one patient receives a heart–lung transplant from a cadaver donor, while his/her original heart is transplanted into another recipient) raise these issues sharply. In this case, the donor of the heart remains alive, leading to people having conflicting feelings about the fate of their heart (Sharp 1995). Sharp's study illustrates how concepts of gift-giving may become understood and used in changing social relationships. Thus, some families of donors may seek out recipients of organs, describing the search as looking for 'their new families'. Whatever the origins of the heart, transplant patients may also believe that they have acquired the characteristics of the donor (Basch 1973; Houser et al. 1992) – claims which have occasionally been sensationalised in the media (for example, Jeffreys 1996). Part of the problem is the cultural heritage of a perception of the heart as the heart of our identity.

Xenotransplantation, as a specific form of transplant surgery using non-human donors, poses the questions of self-identity and rejection particularly strongly. In 1996, a biotechnology company called Imutran announced that it had developed a pig which had some genes derived from humans; as a result, the recipient's body should not reject the donor heart so readily, the scientists reason. Advocates of transplantation from animal donors typically stress the supply shortfall of human hearts, which animal organs could potentially fill.

Emphasis on the mechanical power or failure of the heart is, not surprisingly, prevalent in media accounts of xenotransplantation. The language of biomedicine remains deeply reductionist. But other metaphors coexist with the plumbing, particularly in public statements to the media. The breeding of the first transgenic pigs for possible transplantation provoked much public debate (see for example, 'Ethics of Heartless Pigs': letters to *The Guardian*, 28.8.95). Questions were asked about what it means to be human, on the ethics of breeding animals for such purposes, or on what might happen to the human patient after surgery.

By contrast, the scientific discourse accompanying the new technology minimised concerns about crossing species boundaries and emphasised possibilities of control. One justification made in public debates was that pigs are used for meat; why, therefore, should we not take their hearts into our bodies in some other way? Doctors, moreover, have been using pig insulin and pig heart valves for some time: 'morality cannot be tissue specific', argued Imutran's research director (White, quoted in the *Daily Mail* 13.9.95). The reductionist assumptions are clear; the molecule or tissue does its job and we can ignore the context or control the outcome.

Scientists have long been fascinated by the possibility of moving an organ from one organism to another. Earlier attempts to do so, however, were fraught with problems. Despite the rhetorical insistence on the heart as being 'only' a pump (and the Cartesian heritage of animal bodies as merely mechanism) the mechanism repeatedly broke down in spectacular ways. An organ moved from one species to another rapidly becomes black and swollen as the recipient's body rejects it. Between species, the differences are greater, so the rejection is more spectacular.

These narratives, particularly those focusing on xenotransplantation, are deeply rooted in reductionism and the belief that organs are essentially interchangeable. They ignore the biological context of the tissues; thus advocates of xenotransplantation play down the risks of diseases transmitted from donor animals (zoonotic diseases). They also ignore or play down important cultural meanings and symbolism. Transplantation between humans raises questions about the meaning(s) of death, and corporeality. Thus, contrary to medical belief about the heart as just a pump, lay statements about heart transplants might emphasise cultural symbolism, rendering transplants 'sacrilegious' (Calnan and Williams 1992).

To that we can add all the layers of cultural meaning associated with crossing the boundary between human and animal (Birke 1994; Birke and Michael 1997). The pig, especially, is an animal held by many people to be unclean, and eating its meat is taboo in some cultures. And if some heart transplant recipients feel that they have taken on the characteristics of the donor, what will they feel if the heart came from a pig? But these issues are not particularly relevant to the biomedical construction of our bodily interiority in terms of spaces with inter-changeable organs, or in terms of abstract control systems.

It is at this point that the tension between the metaphors of heart as pump and heart as symbolic seat of the emotions is most poignant. The contrast is very clear between the scientists and doctors, with their language of the heart as mechanical pump, and the other cultural meanings of the heart.

AT THE HEART OF THE MATTER

I have endeavoured to sketch out some of the myriad ways in which metaphors of the heart litter our cultural heritage. Apart from the deep symbolism of the moral/emotional imagery of the heart, there are the equally rich narratives of biomedical discourse. But what matters is not the variety of stories these tell, but the power of these narratives to structure practice. It is precisely the plumbing metaphor that so powerfully constructs how we view heart transplants (or at least, how we are supposed to); that discourse is explicitly mobilised by medical personnel in their dealings with 'heart' patients.

If we have learned culturally to think about our insides as having space, then we can more easily imagine moving bits around in it. It is rather like reorganising the furniture in a room. And we have also inherited concepts of controlled systems, consisting of lengths of pipework and regulatory valves. It is at times useful to think about our insides in these terms, particularly if we are facing heart surgery or if we need drugs such as beta-blockers to control the rhythms of the heart. But it is small wonder that many people find that language uncomfortable.

The heritage of other images always lurks in the shadows. The heart is, culturally, much more than 'just' a pump, carrying with it a plethora of other meanings, and a history deeply entwined with contested power. Among other things, those of us who live in Western industrialised countries learn that we are at 'high' risk of coronary heart disease; we may learn that this is due to our 'high fat' diets, or to the stress of our lifestyles. Through that rhetoric, we learn to understand the heart as an organ over which we have individual responsibility (so ignoring differences by race and social class, as well as gender, in the incidence of heart disease – factors which have rather more to do with social context than individual responsibility: see Farrant and Russell 1985).

To understand the experiences of heart patients, to understand the practices of the clinic or the hospital, we need to unravel all these influences in the context of the intertwined strands of cultural meanings about the inner workings of the body. Unravelling such strands has implications for our theorising. Mechanical metaphors are certainly encouraged by medical personnel (who are heavily invested in the scientific discourses of biomedicine), and by the practices of the clinic or hospital. But they are also encouraged, I suggest, by the theoretical work that so insistently focuses on the exterior of the body – the body as inscriptive surface, the body as a means of performance. To think about the inside, we seem to have to fall back onto the mechanical metaphors of biomedicine. We have also culturally inherited the metaphor of the breakable heart, the seat of emotions; but we do not usually use it to think about what goes on inside our bodily selves. Rather, if we think about it at all, the mechanical model surfaces. Yet is that model not itself a cultural inscription on (or in) the body? It seems to me that we succeed only in perpetuating mind–body dualism if our emerging theories continue to relegate the inside – including the heart – to the realms of biomedicine.

Unravelling the meanings also has significant implications for our experiences of embodiment, particularly for those people who have experienced heart disease in any form. The gap between the medical discourses of the body as machinery, and the experiences of patients with their own perceptions of health/disease and of bodily parts, is immense. Having severe pains in the chest is a terrible experience; it frightens us, and can be the harbinger of disability. This experience is not helped by talk of problems with the plumbing, for such talk fails to address the embodied experiences of people for whom the heart carries much more symbolic power than 'simply a pump'. We cannot, after all, exist without it.

Yet the metaphors remain entwined. Doctors themselves are, of course, heirs to that cultural history; the quotation from Christian Barnard at the beginning of this chapter testifies to his use of more popular metaphors of the heart. Lay people (most patients), too, learn the dispassionate language of the body–machine. To accept the prospect of heart surgery means that one must learn to speak about the heart as a pump (which may not mean accepting that language). For patients, as well as for medical personnel, the clinic or hospital is a site of complex negotiations of meanings; everyone concerned must (re)write their story of their bodily insides. Patients for example must learn how, and when, to refer to the plumbing. In turn, theoretical analyses of the clinic need to pay more heed to the complexities of those negotiations.

Thinking about the entwined metaphors and their implications for clinical practice reminded me of a time, a few years ago, when I was hospitalised with terrible chest pains. When I first experienced the pains, I went to the GP's surgery; I was crying with the pain (and, no doubt, the fear). The doctor who saw me said he could find nothing wrong with my heart or lungs: 'it must be boyfriend trouble', he suggested! Apart from his heterosexist assumptions, he had to conclude that the pain resulted from my emotional heart; the pump appeared to be working. I was in too much pain to get angry and tell him I am a lesbian; I dread to think how the conversation would then have gone.

My scientific training, of course, tells me to think of pumps. At times, I must still speak that language. It tells me, too, to imagine my internal functioning in terms of carefully regulated control systems, represented by boxes and arrows. When I went into hospital with pleurisy (having escaped the clutches of the GP who first saw me), I remembered these boxes and arrows as I watched, in painful fascination, the output of the ECG machine. The clockwork, the systems of regulation, I was told, were all perfectly OK and subject to the normal controls; the pain was not, apparently, emanating from my pump.

Yet my cultural heritage also reminds me that I can lose my heart, can have it broken, can give it. I do not imagine this heart when I open the textbooks, nor did I do so when hooked up to an ECG machine. Rather, I might invoke it when I read poetry or listen to music or make

133

love. This kind of heart relies on narratives that are less controlled (who wants to control the passions of the heart?), and seem to imply possibilities of change, in ways that the mechanical heart does not (unless surgery intervenes). Indeed, possibilities of change are what seem to be missing from so many of the biomedical accounts of how our inner bodies work; the reductionism of physiological systems is centrally about control. The reductionist logic that permits the movement of organs between bodies, or the movement of bits of DNA between individuals, assumes that the parts are fundamentally separable. One part may control another, but they do not seem to be changed within themselves. Such implied stasis, versus other tales of biological processes as change, are themes I explore in the next chapter.

7

The Body Becoming: Change
and Transformation

> imagining is not merely looking or looking at; nor is it taking
> oneself intact into the other. It is, for the purposes of the work,
> *becoming*. (Morrison 1993, p. 4)

The body – to judge by the titles of the many books on the theme – seems endlessly malleable. It can be 'volatile' or 'flexible', it can be 'leaky', it can be made 'slender', it can be 'rejected' or 'deviant'. Upon its surface we can etch the cultural angst of the West at the end of the twentieth century; we can cover it with tattoos, we can pierce its surface folds, we can decorate it with fascinating invention. Yet, as I have argued throughout this book, these shifting meanings are set partly against a backdrop of an internally controlled inner body, the body of physiology. This body is carefully regulated, almost against our will, and seems to change little.

Previous chapters of this book focused on the kinds of models and metaphors predominant in thinking about, and representing in diagrams, the physiological body. I have concentrated on how these models prioritise control and regulation, on systems and mechanisms. Insofar as these concepts enter social theorising, they do so in two ways: as the foundation for rejection of 'naturalistic' views of the body as biologically determined, and (contrarily) as metaphors extending, through information and systems theory, to the potentially transgressive worlds of cyberspace (see Tomas 1996).

Yet both these views, in different ways, are reductionist, and contribute to the underlying problem of seeing the biological body as bedrock. Such conceptualisation of 'the biological' never escapes determinism; the biology is always constraining, unchanging. In the first case, the biological body is simply assumed in its constancy. In the case of the 'cyber' literature, the body simply disappears. Real human bodies in computer culture become 'meat', from which virtual reality allows one to escape like a pure distillate (Lupton 1996; Tomas 1996). Here, the materiality of the body is simply ignored, or seen as something to be transcended as we enter virtual worlds. But, I would argue, neither of these views escapes the problem of seeing biology as constraining; rather, the biological body is simply taken for granted, and implicitly ignored, in both. And it is precisely in its taken-for-grantedness that the fixity lies.

In this chapter, my concern is to move away from notions of fixity and constraint, and of the body as given, towards an understanding of the biological body which allows for dynamic process. That understanding is one of the body *becoming*, as transformative. I want to question the notion of the biological body as bedrock. I do so partly because I consider that it gives a limited understanding *of* our biological bodies, and partly for more political reasons, because such a view of biology leads to determinism.

Before beginning this journey, however, I want briefly to revisit some themes from recent feminist and social theory regarding the body. Malleability and agency seem to be key themes, while simultaneously assuming the biological body as imposing constraint. The notion of biology as constraint is shunned in such theorising, because it seems to imply determinism. But whatever goes on the level of 'the biological' need not be so, as I will argue.

The discourses of biomedicine have undoubtedly contributed to a perception of fixity and determinism – the emphasis on control in physiology is one example; yet there are alternative views, other stories to tell, from within biomedicine itself. It is these which are my focus here. In seeking to tell other stories, however, I am not claiming that some are more free of biases or metaphoric structure. On the contrary, no stories that we tell about nature can ever be innocent, as Donna Haraway cogently reminds us (Haraway 1991c; see also Harding 1991). But some can be 'better' stories, in the multiple senses of being more faithful accounts of biology, and of being more sensitive to the political grounding and context in which knowledge is created.

In the quote beginning this chapter, Toni Morrison is writing about the experience of a black woman writer, herself writing in a racialised and genderised world. Imagining, for the writer, must necessarily entail becoming. I use that quote here, in quite another context, to emphasise both the 'becoming' of the body as transformation and the 'becoming' implied by Morrison in the related acts of reading and writing. Telling 'better' stories is part of that process of imagining a less reductionist biology. Theories presenting alternatives to reductionism may still be stories grounded in a particular sociopolitical context – but they can also be better stories for feminism because they begin to move us away from reductionism and determinism, with all the attendant problems. They might also be better stories in the sense of being more faithful accounts of the biological processes themselves.

THE BODY REACTS BACK?

One of my premises in this book is that a perception of the biological body as synonymous with mechanism and fixity underlies its absence from much of feminist theory. Rather, body theorists have concentrated on the socially constructed body – the malleable surface of an internally stable corporeality. Much of this work on the body has drawn extensively on the work of Foucault, emphasising particularly the ways in which the docile or disciplined body is created discursively. This is particularly clear in feminist work analysing the disciplining of the body in, for example, anorexia or bodybuilding. But however useful Foucauldian analyses have proved, there are also ways in which they are limited. In particular, it can be argued that the body, for all its apparent centrality in Foucault's work, actually disappears as a material entity (Turner 1984; Shilling 1993).

To put that another way, 'The body is affected by discourse, but we get little sense of the body reacting back and affecting discourse' (Shilling, ibid., p. 81). The biological body thus disappears, constructed in such work only as and through discourse but not the reverse. This discursive body might be contrasted in social theory to concepts based on the agency of the body, while the biological body remains a silent shadow.

In Chapter 5, I examined some of the predominant narratives of physiology, which are still those of control, of systems, of mechanical regulation. This is how we have learned to speak of the inner workings of the body, to construct our understandings of the body (drawn from other cultural contexts) and to impose these *on* the body. There is also a sense in which the body does 'react back', to use Shilling's term – for the material body itself imposes constraints on our understanding.

The body surface, however, has in the West become a project to be worked upon, 'an entity which is in the process of becoming; a *project* which should be worked at and accomplished as part of an *individual's* self-identity' (Shilling 1993, p. 5; emphasis in original). This body is always superficially transformable. Various practices illustrate this: bodybuilding, body piercing, anorexia, cosmetic surgery, transsexual surgery, even cosmetic surgery as performance in the work of French performance artist Orlan (Davis 1997a).

We can thus 'make over' the body in various ways;[51] yet a shadow remains. The internal, anatomical body seems to me to be haunted by the ghosts of fixity and constraint. Admittedly, we can 'tune up' some parts of our anatomy – the cardiac workout for the heart, or exposing ourselves to infection for the immune system, for example. But 'the biological' always seems to be foundational, the ever-present bedrock to our theorising. Thus, sociologist Bryan Turner for example speaks of the body as potentiality; yet in the same volume he notes the attempt in contemporary sociology to 'work out a systematic bridge between the voluntaristic theory of action and the biological foundations of the human organism (as a set of constraints)' (Turner 1992, p. 81).

It is that assumption of the biological body as a set of constraints that I take issue with. That is not to say that I am denying that there are constraints – human bodies cannot fly, nor can we breathe underwater, for instance. What I seek to challenge is some of the assumptions embedded in that claim.[52] Whether Turner intended it or not, 'constraints' seems to connote fixity, the body as the bedrock imposing limits on what the human might seek to do. The 'science of action' (as Turner (1992, p. 162) described sociology) thus seems to rest on a science of constraint, even of inaction at the bodily level.

Indeed, it is that underlying vision of fixity that has contributed to what might be called sociology's flight from the body, for fixity all too easily becomes equated with essentialism (see Fuss 1989). Yet insofar as the flesh of the biological body offers constraints to our possibilities, these are dynamically generated; they do not necessarily imply fixity. It is such alternative narratives – of possibilities, not determinisms – that I want to explore here.

REDUCTIONISM REVISITED

Biologists often seem quite irritated by anti-reductionist critiques of science – whether generated by feminists or anyone else. We know, they aver, that genes for example do not act in isolation, do not 'simply' code 'for' some complex trait. In offering critiques, I have frequently been confronted by irate scientists, who seem to believe that 'no one really believes that' or does not believe it 'any more'. It is as though I am somehow a traitor – a biologist who perversely insists on exposing the soft underbelly of biological thought. But perhaps I have set up a straw man/woman; perhaps I – and other feminist biologists – have overstated the case?

Or perhaps not. Certainly, there are some voices among the scientific community who speak publicly and overtly of genes determining; Richard Dawkins is a familiar example, with his narratives of selfish genes intent on pursuing their own ends – narratives in which the body is entirely irrelevant except as temporary home for the genes. Such voices are well-publicised, and it is much harder to hear the whispers of dissent.

Moreover, however much biologists may claim that they are 'inter-actionists', reductionism remains hegemonic and global. Indeed, biologists are sometimes rather too glib in their acceptance of the idea of 'interaction of genes and environment': the phrase, remarks Steven Rose, 'masks as much as it reveals' (Rose 1997a, p. 131), since both terms are problematic and highly abstract. What is interacting, and how? What, exactly, is 'a gene' (a highly contested term, even among biologists)? Reiterating 'interaction' does little to challenge the hege-monic stories of reductionism.

Alongside the reductionist story, however, there are dissenting voices and other narratives. It is quite true that biologists know perfectly well that genes do not act in isolation. Yet the story that they do, and that they are primary, has enormous power. Belief in the 'master molecule' story of DNA persists, notes Donna Haraway – indeed, it dominates throughout our culture. The knowledge that this is not an accurate picture of DNA 'is entirely orthodox in biology, a fact that makes "selfish gene" or "master molecule" discourse symptomatic of something amiss at a level that might as well be called "unconscious"' (Haraway 1997, p. 145). How – if biologists themselves scorn the master molecule story – has that tale become so culturally predominant?

Haraway discusses this in terms of 'gene fetishism', though the problem of reductionism extends well beyond the reduction to DNA. There is a profound cultural reliance on stories that reduce the complexities of the biological body (or of biology more generally) to component parts, to a narrative of unchanging essence. While I, and many others, have written before of alternative stories, we must go on elucidating them – as I must do here as an antidote to the visions of physiological fixity I noted in previous chapters.

There is no doubt that the reductionist programme has been

successful; it has generated enormous amounts of research, led to predictions, and enabled significant developments in biomedicine. It has also (and significantly) generated enormous amounts of money, notably through the recent development of patents for biological material.

At the beginning of the twentieth century, reductionism battled against more descriptive accounts within biology. But reductionism was seen as 'more scientific', in that it assumed experimental approaches and sought explanations in physicochemical processes. Thus it eventually won out. Physiology, for example, had long adopted an experimentalist and reductionist account, focusing on the maintenance of the organism. By contrast, those approaches that insisted on understanding how organisms develop – embryology, for example, and some approaches to immunology (Tauber 1994) – declined.

Reductionism as a way of explaining or understanding is, moreover, facilitated by the methods of science. To exclude influences other than whatever it is we study, we must control conditions in experiments; that process of control, of careful delineation of an experimental design, is itself reductionist for it encourages us to seek explanations that are themselves prioritised by the experimental design itself (see Birke 1986; Rose 1997a). Furthermore, the very dynamism and vitality of the living organisms that we study in biology is lost and destroyed in the processes of our investigation. This in turn connects to the abstraction and reductionism of the scientific diagrams that I outlined earlier; for in the process of abstraction, we 'freeze' the representation. Dynamic processes are thus lost in the representations of diagrams.

To study mammalian physiology, furthermore, usually means killing the animal, so losing its vitality for ever. To open up the living body is to subject it to brutality. As we study the dynamic processes of life we fix them into the eternal stillness of death. And at the more microscopic level of the cell, the active processes become lost through the 'brutalising techniques of the electron microscope' (Rose 1997a, p. 169). We can understand life only at the moment of our destruction of life, it would seem.[53]

Whatever the successes of reductionism, much has thus been lost in the stories that we might tell of how our biology works. In his analysis of the history of ideas in immunology, Tauber points out how reductionism remains ascendent, despite the various attempts to rescue the whole organism from reduction to constituent parts. The problem, he notes, is that all we can do is to indicate our appreciation that nature is more complex and pluralistic; 'biologists have yet to erect the scaffold of an alternative 'new biology', where laws governing complex hier-archical systems cannot rely solely on characterising isolated phenomena ... The experimental power of the reductionist program has driven the study of biological phenomena to the genetic and molecular levels of investigation, but in the process of establishing its hegemony, the holistic basis of an earlier biology was lost' (Tauber 1994, p. 54).

THE TUNED UP BODY: IMMUNE INFORMATION ACQUIRED?

Yet if determinism and the power and agency of the gene are undoubt-edly predominant motifs in modern biology, they are not uncontested. Alongside these powerful stories are also narratives opposed to reduct-ionism, ones based more on organisms than on genes. There are, for example, views emphasising the cooperativeness of organisms and of parts of organisms/ecosystems; there are narratives within biology which prioritise the dynamically evolving organism, constantly respond-ing to change and changing itself in the process. Neither of these are reductive, and both are centred on the organism and its potential for change (Tauber 1994, pp. 57–8). In the next two sections, I want to explore two of these 'alternative narratives'. The immune system is one area where alternatives to reductionism seem to be developing; but, I argue, there remain problems, particularly with the concept of information embedded in immunology.

Change and flexibility are becoming key concepts in research on the immune system, for immunity can be acquired, even worked at (Martin 1994). And, partly because of the sometimes gendered ways that the immune system is described and because of the growth of interest in immunity since the advent of AIDS, the immune system has received feminist attention (Martin, ibid.; Haraway 1991b).

Perhaps not surprisingly, the cells of the immune system are marked by gender, race and class in biomedical accounts (Martin 1994, p. 59). Martin illustrates this point with examples drawn from popul-arisations of how the immune system works; thus, we meet 'Bubbles', the B cell dancing about in the blood, while macrophages sometimes become 'feminised housekeepers' in these tales (ibid., p. 56). That these are popularisations does not invalidate her point: the popular images gain widespread cultural acceptance, and must reflect implicit assumptions made in the scientific literature from which they derive.

The immune system, moreover, has become a truly postmodern system of the body, transcending the old order of mechanistic

physiology. The immune system centres on the metaphor of information; it is now a 'polymorphous object of belief, knowledge, and
practice' (Haraway 1991b, p. 204). The biological body, suggests
Haraway, is now no longer to be understood in terms of the homeostatic systems of earlier physiologies, but has become a body 'based on
codes, dispersal and networking, and the fragmented postmodern
subject' (ibid., p. 211). I disagree with her belief, as stated here, that
homeostatic systems are now superceded: rather, they seem to me to
persist alongside her vision of the 'postmodern' in biology. But I want
to emphasise her point that the immune system is the site of a marked
discursive shift.

There have been a number of such shifts in narratives of the
immune system. Perhaps most significant has been the change from a
language of strategic defences aimed at keeping out invaders, to that of
a more responsive system which can be 'trained'. This represents a
shift from seeing immunity as a largely passive process, to seeing it as
active and specific (Martin 1994, pp. 33–4). The body no longer has
the integrity of the fortress (as earlier narratives implied), but is
dispersed into flows of information – a postmodern subject, indeed.

With the much greater public profile of the immune system in these
days of AIDS, a central motif of immune system theory – the distinction between 'self' and 'non-self' – has gained much wider currency.
The immune system, in order to 'attack' invading cells such as
bacteria, must learn to distinguish between cells that belong to its own
body ('self'), and those that do not ('non-self') and which threaten
bodily health. Thus, the 'non-self' is categorised as potentially hostile,
as other to the self (see also Tauber (1994) for an explanation of the
cultural and philosophical contexts of this separation of self from non-
self).

Although there are alternative views, for example, that the immune
system cannot distinguish 'non-self', only myriad forms of 'self'
(Martin 1994, p. 110), or that the immune system is responding to
'danger signals' (Pennisi 1996), the self–non-self distinction remains
central. And it gives rise to the notion that it is possible to 'work' on
one's immunity by giving the immune system 'practice' at dealing
with non-self. In Martin's study of perceptions of immunity among
various communities in the Baltimore area of the United States, she
notes an emerging discourse of what she terms 'immune machismo'
(ibid., p. 236). My immune system is strong; yours is weak. I can
strengthen mine, so making my-self stronger.

This rhetoric, she argues, parallels the language of 'flexibility' in corporate management. Both good managers and good immune systems must be flexible, able to respond to a variety of situations. Those who are insufficiently flexible will go to the wall. As Martin notes, we might welcome the breakdown in the old order of exploitation of workers by management; but we are also witnessing a new form of discriminatory practice, one built upon 'flexibility'. At the very least, the management training courses to which Martin refers must inevitably discriminate against those who are, say, pregnant or disabled, for they rely on ideas of 'tuning up' the mind through tuning up the physical body, by means of a kind of assault course training. Only those who survive the gruelling tests will be tough enough for management.

Emphasising the changeability of the immune body appears to challenge notions of the body as fixed or determined. Yet it does so in ways that I find troubling. Apart from the way that discrimination is quite literally embodied in management training courses, the new language has emerged within a wider cultural context of corporate capitalism (for all its apparently welcome emphasis on flexibility and change). It is perhaps unsurprising that 'immune machismo' results. Indeed, notes Martin, the language of the immune system generates a new form of social Darwinism, within which some immune systems are seen to be better than others. Those with weakened immune systems – whether they inherited that trait or acquired it through, say, AIDS – are somehow less 'fit'.

Indeed, that point reminds me of my own bodily limitations, in late middle age. The 'privileged pathology' of the late twentieth century, suggests Haraway, is stress – communications breakdown (1991a). Was it the 'stress' of overwork that made me ill with chronic fatigue a few years ago? Or had I not done enough to tune up my immune system to enable me to fight off with virile strength the virus that preceded fatigue? Was developing chronic fatigue a sign of my own feminine weakness, an inadequate immune response?

Certainly, like Susan Wendell (1996), I have subsequently met both sympathy and intolerance; even in liberal (?) academia there remains more than a trace of machismo. Couldn't I take the pace – perhaps the illness was somehow my fault? Had I planned my career badly, or failed in my time management? Where once it was the menopause that was said to show women's weakness, now it is illnesses like chronic fatigue which show up your/my failure to deal with the system. Just as gene fetishism puts blame for my failures squarely on my DNA, so the new immune rhetoric makes me personally responsible for my failure to stay healthy. In doing so, of course, it shifts attention away from any possible environmental influences. Given how much we are all exposed to a cocktail of chemicals, as well as inhumane work organisation, that rhetorical shift is rather handy.

Another problem with the 'new' immune discourse is that it seems to play into the loss of the body in social and cultural theory. Information flows are all, and we thereby lose any sense of the organism itself. Haraway, for example, emphasises the shift from narratives of master control (that dominated biology up until the 1970s, she argues) to 'postmodern' narratives (particularly in immunology). There are, she suggests, important implications of the shift she describes for how we understand – and hence experience – the body and its changing states in health and illness. The concept of the organism, as an entity, has broken down, she argues, replaced with flows of information:

> Organisms are made; they are constructs of a world-changing kind. The constructions of an organism's boundaries, the job of the discourses of immunology, are particularly potent mediators of the experiences of sickness and death for industrial and post-industrial people. (Haraway 1991b, p. 208)

Haraway is, I believe, right in recognising the (re)construction of the organism in terms of coding and information. Within biology as a whole, students now must learn about information flows and their mapping; it is genome projects, not organisms, that mark the centre of biological knowledge. This is not a move I welcome, for both biological and political reasons.

My first reason for being troubled by Haraway's account has to do with biology. She is surely right in pointing out the power of the information metaphor, and the ways in which it contributes to the literal reconstruction of organisms through genetic engineering. But the problem with the information metaphor is that it seems to imply that 'anything goes'; information flows in and out of the (permeable) body or the genome. Yet organisms are not so fluid: rather, their processes are self-organising in ways that generate constraints (see below).

That is not to say that Haraway is herself arguing for the information model of technoscience. She does not. What troubles me is the way that her work is often read as implying a kind of postmodern free-for-all in perceptions of biology (or even as having little to do with *biology* at all[54]). The critique and the possibilities of other kinds of narratives in biology seem too often to drop out of debate. Consider, for example, her insistence in the extract above that it is the job of the *discourses* of immunology to construct the organism's boundaries. Indeed they do; for out of the self/non-self distinction of modern immunology we come to understand what counts as me or not me. My boundaries are thus conceptually constructed.

But they are more than that. Haraway is, I believe, too easily read in ways that lose the materiality of the organism. It is not only the discourses of immunology and culture that construct my boundaries, but also the various cells that are busily making and remaking my tissues. I can, perhaps, escape the 'meat' whenever I explore the Web or write on my computer; meanwhile, the meat is busily reconstructing itself, an unseen materiality.

Susan Wendell (1996) makes a similar point regarding Haraway's work, in her analysis of disability. She is concerned that emphasising the cultural constructions of the body so easily loses lived experience, as 'though discourse and its political context are all there is, without acknowledging either the reality of physical suffering (for example, by people with AIDS, ME,[55] MS, Amyotrophic Lateral Sclerosis (ALS), rheumatoid arthritis), which surely has *some* relationship to the development of immune system discourse, or the effects of this discourse on the lives of people who are thought to be suffering from immune disorders' (Wendell, ibid., p. 44). The organism or body, suffering or not, seems to slip out of many postmodern accounts.[56]

Haraway's vision of the cyborg (1991a) similarly implies fluidity and loss of the organism. She speaks of 'polymorphous, information' systems, emphasising rates of flow across boundaries rather than bodily integrity. Haraway opposes holistic/organismic views (and related stories of development as progress), as fostering a kind of solipsism. In her utopia, organisms seem to disappear into webs of complexity, with entities dispersing into information; they become 'strategic assemblages ... ontologically contingent constructs' (1991, p. 220).

By contrast, I want to hold on to 'the organism' as an entity. Partly, this is for biological reasons, which I will deal with below, and partly for political reasons (for further critique of Haraway and the cyborg image, see Stabile 1994). Insisting on information flows rather than organisms supports the reductionist view of organisms as mere artifacts of the genes. Information and fluidity seem to deny the fleshly boundaries of the organism, as bits of information (coded in DNA or in the action potentials of neural functioning) flow in and out.

In the highly commercialised practices of genetic engineering, segments of DNA are moved around and 'new' organisms created.

These in turn can be patented, for profit. The borders between organisms – what constitutes you or me – seem to dissolve, no longer to matter. This seems to be the technoscience version of the post-modern narratives in which organisms become merely information flow.

But this is to ignore the complexities of living organisms, in which bits of DNA play only a part. One consequence of such denial – often noted by environmental activists – is that we run a terrible risk that genetically modified organisms (a cereal crop, say) escape from 'proper' controls and introduce unwanted characteristics into the wider environment. Disease-causing bacteria could thus escape, or we might witness rampant destruction of an ecosystem as the usual biological controls break down following deliberate genetic manipulations (see Newman 1995).

Denying or downplaying the integrity of the organism – as the genetic/information model does – permits us literally to dismember it. So, we can alter pigs genetically, giving them 'human' genes, so that we can remove their hearts for human transplant surgery. The conceptual dismembering of reductionism enables the literal dismembering and death of the living animal. Cruelty, and a whole range of oppressions, stem from that denial of integrity (see Birke 1994).

These, then, are the reasons why I find the 'information' model so discomfiting. I must acknowledge, however, that it is also very powerful, both in biomedicine itself and (in different ways) in cultural critiques such as feminist science studies. The hegemonic gene story is told and retold in feminist work; there is endless fascination with the mysterious gene and its mapping. But, even as it is criticised, it acquires new strengths, and the gene/information story becomes hegemonic in feminist critiques of science themselves.

My discomfort does not mean that I cannot recognise the advantages of the 'information' model in challenging reductionism. The narratives of the immune system as told by Haraway and others remind us of the flexibility of bodily systems, and of the impossibility of separating our insides from what goes on outside, or of separating that internal body from our culturally mediated understanding of it. In that sense, the information model challenges the centrality of the individual – and hence of individualism – through its insistence on information flows across boundaries. By contrast, exchange between inner and outer tends to disappear from reductionist accounts.

But the other side of the coin is that the open-ended flexibility of information flow is itself reductionist, precisely because it reduces 'information' to simple molecules like DNA, thus prioritising them over other components or processes of cell, tissue or organ. And as it does so it feeds into the crass and dangerous assumption that we can easily move bits of DNA around in genetic engineering, without thought or consequence. The primacy of the moveable gene in bio-medical discourse and practice runs parallel to its primacy in the leaky information flows of postmodern narratives.

Reductionism, then, appears in many guises. And it has many critics. I want now to pick up on two, intertwined, problems of reductionist approaches within biology. In doing so, I am drawing on what I believe to be 'better' accounts of how biological organisms work. By 'better' I mean narratives that seem to fit more closely the ongoing processes of life (see also Haraway 1991c; Harding 1991). However hegemonic and culturally powerful reductionist stories have become, there are alternative tales on which we might draw, even within the discourses of biomedicine. Might these help us to escape the problems of seeing the biological as bedrock, as determining?

THE ORDERED BODY: ORGANISING SELVES

One problem with reductionism is that, in focusing on molecular explanations, we lose any sense of the organism as a whole. Indeed, as genetic and molecular reductionism have become so dominant, the organism has largely disappeared from the discourses of biology (see Goodwin 1994). Where once the phenotype – the bodily and behav-ioural characteristics of the organism – was preeminent, now it is the genotype – the sum total of the genes. Analysing the historical development of the conceptual rupture between phenotype and genotype, Gabriel Gudding notes how that split 'led to a sense of the body's fragility and its eventual "disappearance" as a seat of agency, morality and identity' (1996, p. 525). Detonation of the atomic bomb in mid-century only served to underline the body's fragility as well as to reinforce the genetic narratives, he notes; for mutations caused by radiation are unseen, yet can be lethal.

From that fragility, the phenotypic body has now almost entirely disappeared. It has lost agency, which is now ascribed instead to the gene. Indeed, as Gudding notes, human bodies might even now be read as 'genetic stutter' – almost literally. He quotes a description of Nancy Wexler, who has worked for years to locate the gene involved in Huntington's disease;[57] working to isolate bits of DNA means learning to 'read' the stripey patterns produced onto gels once DNA

147

has been extracted and broken down. When walking in Venezuela, she begins to see 'the shadows cast by a venetian blind [on people's faces]. Everywhere she looked, the tri-nucleotide stutter looked back, dizzying in its persistent misery' (ibid., p. 528).

Seeing the patterns of DNA tests on the faces of people may seem a strange fantasy; yet it is an imaginative consequence of the loss of the phenotypic body from the narratives of biology. The language of the genotype, the metaphysical and literal power of the gene, predominate. It is read onto the body, while the body itself, the whole organism, disappears from the tales told by biology. At best, they remain shadows frequenting the margins.

Perhaps part of the reason for the success of the genetic programme story is that, for all that modern genetics may seem very secular, it draws on theological notions of creation. Stuart Newman (1995) notes how Darwin, in developing his theory of evolution, relied on prevailing beliefs in the necessity of divine creation. While his successors disavowed the direct link to God, they still needed to appeal to some external force which directed evolutionary change, which would 'give form and reproducibility to the inert materials of their biological world. This force was soon provided by the theory of the isolation and continuity of the germ plasm ... and its eventual recasting into the scientifically questionable modern idea of the "genetic program"' (ibid., p. 213). What created bodily form, then, was not organisms themselves; but nor was it any longer God – it became instead the genetic blueprints of DNA.

The study of biology may seem to be the study of living organisms (or perhaps more accurately the study of dead ones). Yet, as Brian Goodwin (1996; and see 1994) has pointed out with regard to the rise of molecular biology and genetics, there is no theory of the organism as a self-organising, dynamic, transforming entity, in the way that biology is currently taught. Organisms are simply epiphenomena, accidental byproducts of genetic plans. Yet, Goodwin argues, the reductionist strategy emphasising the primacy of the gene fails to tell us much *about* the organism, including, I would add, our own physiological selves.

Relatedly, reductionism does not adequately explain how bodily form arises. To the genetic reductionist, the form of an organism is the direct result of information coded in the genes. But this is not enough to *explain* completely how that form arises. We share 99 per cent of our DNA with our closest relatives, yet our bodily form is rather different from that of the chimpanzee.

The genetic story also fails to address the changes in an organism throughout its lifespan. We humans go from fertilised egg to embryo to foetus, through childhood and puberty to adulthood and then to ageing. Some organisms undergo even more dramatic changes; the metamorphosis of tadpole to frog or from caterpillar to butterfly are clear examples. Various body forms – yet the same DNA. We can, of course, speak in terms of some genes being switched on or off at particular life stages. But that is clearly not enough. How does this happen at that particular time? And why? Why are some developmental processes fairly stable, while others seem to be susceptible to change (there are, for instance, certain periods of development before birth in which the developing foetus is particularly susceptible to, say, chemical toxins)?

Constructing accounts of biological processes in terms of 'information' ignores their complexity. Cells simply do not work solely by reading off 'information' from DNA. Steven Rose makes this point using the analogy of the cassette player. The information metaphor implies that DNA is read much like a cassette tape. But, he points out, cells and DNA make and remake themselves in their mutual engagement – quite unlike the passive reading of a tape (1997a, p. 130). 'Information' is almost a consequence of that engagement rather than its cause. Rose insists on speaking of the developmental trajectories of organisms – what he calls their 'lifelines' – to emphasise the ongoing process of engagement of the organism in its own life history.

Rose also questions the assumptions implicit in the widespread belief that genes are preserved across generations (at least if they are useful ones). Preservation and transmission of genes is crucial to the rhetoric of neo-Darwinism. And the idea has become a familiar one in popular culture, as we ask ourselves what genes we are passing on to our children. Yet what exactly *is* being passed on? It is not a particular chunk of DNA itself, as that has been broken down and copied many times during our lifetimes (and some changes made in the process); thus, the actual composition of DNA does not persist but is, rather, made and remade.

What is implied, Rose argues, by the idea of the 'preservation of genes' is the replication of form, not the persistence of DNA itself (ibid., p. 213). That is, giraffes produce babies that look like small versions of themselves – their bodily form as giraffes is replicated. It is the shape, the form, not the chemicals, that reappears in each generation, yet it is this that is invoked when biologists refer to 'preservation of genes'.

But how is shape conserved? If giraffes are always giraffe-shaped, how does that constancy arise, if not the genes? To many critics of the reductionist story, it is not genes alone which determine the outcome; rather, as the organism develops, its cells themselves create what Goodwin calls morphogenetic fields. These might be electromagnetic or chemical influences that help to create spatial effects, so guiding the movements of cells. How cells move to create organs is thus not directly the result of a blueprint, but is a process that emerges from the prior creation of other cells and structures, to create a new level of order.

Genetic determinism assumes that the development of organisms follows from a blueprint laid down in the genes; the form of the organism is, as it were, preformed and the animal or plant develops passively (Oyama 1985; Birke 1986). But crucial to the development of embryos (human or otherwise) is active engagement, in which the developing embryo helps to shape and interact with its own environment.[58] This conceptualisation, moreover, is the antithesis of the image of the passive or passenger embryo which forms the focus of much feminist criticism.

Yet we are limited by the heritage of scientific reductionism, by the cultural predominance of ideas based on information flow or control systems. As a result, we simply do not have the language in which to describe the activity of embryos in their own self-creation (or at least not adequately). I could write, for instance, about how sheets of cells in early embryos fold over and turn inward, so creating multiple layers. I could say that this is the beginning of organogenesis, the formation of organs, and that each layer influences the movement of the next. But somehow, even as I begin to do so, the words do not adequately convey the concept of self-organisation. It is rather like trying to describe something which deeply moves us; words simply fail.

The poverty of language aside, those movements of sheets of cells are clearly not random. But nor are they simply the product of a genetic masterplan. As cells move, they both help to create and respond to, electromagnetic or chemical fields. Moreover, as the embryo develops, these fields give rise to chemical and physical gradients in which, for example, the concentration of a particular chemical is greater at point A than point B. Certain cells might then follow the gradient, to end up at A rather than B. If there is some degree of inevitability in these processes, then it is because 'this is what cells do'; as each structure forms, it facilitates the formation of another structure. Any inevitability is not because of dictatorial DNA. It can, moreover, be understood only at the level of the cell, not in terms of its constituent molecules. The processes are dynamic, and self-organising. The way that genes can influence these processes is at least partly through altering these fields of electrical, magnetic or chemical gradients.[59]

In this version of embryological events, there is action indeed. But it is an action that disappears in constructions of the organism as fixed or passive. Adults, of course, may appear to be already formed, those formative fields no longer influential. But to assume that is to fall back into fixity. On the contrary, our bodies are constantly being made and remade; bones, muscle, connective tissue – all are constantly in flux.

Such a view insists on seeing organisms – and their 'biology' as transformative and in which the parts are generated *by* the wholes. Form thus emerges out of complex processes not coding in genes. In particular, this view refuses any simple collapse onto genetic determinism, and relocates the organism (and hence the biological body) back from the margins. It also insists on the creation of new levels of organisation, which is a different move from insisting on 'interaction' of genes and environment as biologists sometimes do. Genes are only part of the story.

In turn, what this implies is that such 'constraints' as exist become built in as a result of active processes; they are not simply determining. Rather, we might think of such constraints as constantly in the process of becoming. If, for example, nearly all vertebrates have similar structure of the front limb (ending in a five-digit structure, even if this is often vestigial, as it is in the hoofed mammals), then this need not be determined solely by genes. Rather, it can be the product of emergent processes. As parts of the embryonic limb develop, they influence other parts directly, which in turn exert further influence over neighbouring tissues. To be sure, genes are part – but only part – of this story, as some genes act to switch on or off particular parts of this overall process of development: but they act in context, within the dynamic changes of development in the organism. It is this, active and internally cooperative[60] process which generates the emergent form of an organism, not genes acting as blueprints (Goodwin 1994).

Organisms in such a view are more than just strategic assemblages of cells/information: they are self-actualising agents. Insisting on organisms as entities/agents returns them conceptually to the study of biology – from which whole organisms have almost disappeared in the world of genes as prime movers. If, as Donna Haraway insists, it's problematic to think of organisms in terms of a path of (genetic) progress (a narrative deeply embedded in Western culture), then we should certainly use other metaphors. But escaping into webs of 'polymorphous information' will not do; while these may be powerful narratives of postmodernity (played out in the practices of bio-technology), to lose the integrity of the organism altogether is a highly dangerous move – a theme to which I will return in the final chapter.[61]

There are, of course, problems in insisting on the agency of organisms' bodies in their own development. In particular, it can lead to reinforcement of individualism, especially if what is emphasised is the agency *of* the individual body itself. In turn that could feed into a view of the developing foetus as individual – a view that is dangerous to feminist demands for reproductive and abortion rights.

Yet focusing on organismic agency and transformation need not lead down that path. It will lead to individualism, certainly, if the agency is understood to come *from* within the individual – almost as another version of the genetic programme. But what I seek to emphasise here is an understanding of agency that emerges out of the engagement of the organism with its surroundings; it is thus an agency in relation, not an essential property of the individual. In that sense, the 'agency' of the tissues of the developing foetus can be understood only in relation to the foetus's engagement with its environment – which must include the mother's body.

Ascribing agency, self-organisation and transformativity to organisms/bodies *in relation* works against reductionism and constructions of the biological body as fixed, I would argue; rather, these ideas emphasise the possibilities of change, both from day-to-day and over our lifespans.[62] To see organisms/bodies as having agency both within themselves and in relation, and the ability to be self-organising, also implies that social constructions and experiences of gender can themselves be part of process. 'Sex' cannot thus be prior to gender, but itself shaped by, and contingent to, gender. Put another way, processes involved in creating and continually recreating (sexed) bodies are partly material and partly social/experiential. I have written that sentence implying two sets of components; yet that is itself a problem, for there is no separation.

Meanwhile, our internal organs and tissues also make and remake themselves. Physiological language may be deeply mechanistic at times – it speaks, after all, of control systems, and feedback loops serving to stabilise them, as we have seen. Yet implicit in these systems is *active* response to change and contingency, bodily interiors that constantly react to change inside or out, and act upon the world. Biologists, in our training, have usually learned to read the physiological/cybernetic models as implying dynamic equilibrium.

In principle, the active process of homeostasis is assumed within the feedback models I learned in my training. Yet, I would argue, it is constancy and stabilisation that emerges out of these processes which is the strongest cultural message. Indeed, the metaphor of homeostasis implies a fixed set point – of body temperature, say – rather than dynamic process (Rose 1997a, p. 17). And it is that message of constancy which, I suggest, has become culturally predominant. In part, it is sustained by the deep commitment to normalisation within the theories of biomedicine; stability equals normality equals health. Instabilities threaten chaos, disease, and death.

The concept of physiological systems does allow for some degree of influence from the outside; such 'external' factors as temperature or day length, for example, might be acknowledged in physiology textbooks as influencing cycles of hormonal feedback. But the emphasis on homeostasis tends to imply a buffering *against* the environment, while the body itself seems hermetically sealed. Any concept of the body/organism as either changing or open seems to drop out of the notion of (separable) organ systems.

It is that picture that, I believe, carries over into wider domains. The body of sociological theory is, at its core, stable; it is always, already, an adult body (not an ageing one, nor a prepubertal one), and it is largely a closed system. Thus, changes through the lifespan are not central to this body (adolescence is theorised by those interested in youth culture; the changes of puberty are left to the dusty tomes of biomedicine). Rather, the body of recent theory seems to me to have all sorts of cultural representations written onto its surface, an endlessly changing kaleidoscope of images. But underneath?

Underneath, the biological processes go on. We can act to change the body in certain ways, by working on it, but we do not expect to change the deep structures of the body or its internal workings. We are surprised and frightened when those structures make their presence felt. As Jackie Stacey has noted in her study of cancer, the body betrays, it is out of control (Stacey 1997). Indeed, metaphors of the sick body rely on images of a nature uncontrolled, devouring us from within.

It is only when things 'go wrong' in disease, when things are 'out of control' that we become aware of the processes of the body. That is the only context in which we are likely to think of the body in terms of day-to-day change. And that we do so is itself testimony to the underlying assumption of stasis, which in turn facilitates the determinism feminists must decry.

REMAINING UNCERTAINTIES

To rehabilitate the biological body without resorting to stasis and determinism is an urgent task, I believe. Theorising embodied experience while leaving the biological body to get on with its own work merely perpetuates old dualisms. Like virtual reality, body theory all too often 'leaves the meat behind'. And in that regard, such body theory is just like genetic determinism which philosopher Mary Midgley has described as characterised by a vision of life that gets away from the organic (Midgley 1992[63]). Both seem to embody (if you will forgive the pun) that desire to escape the body and its messy constraints altogether.

Refusing to relegate 'the biological' to categories of fixity helps to move us, I would argue, in the direction of integrating social theory with an understanding of the material body. To that end, I have insisted, in my work on feminism and biology, on transformation and change, rather than fixity and determinism (see Birke 1986, 1989, 1998). That is not the whole story, of course; all the stories we might tell are products of the same techno-culture. All are generated out of social frameworks, embodying the dominant social values in varying ways (Longino 1990). Insisting on complexity does not solve all the problems implied by that heritage. But all the stories are not equal; some are, I believe, 'better' – both in the sense of being truer to my (feminist) readings of 'how nature works' and better in the sense of being closer to feminist understandings of the social/political world.

Yet in thinking about feminist work on embodiment in relation to scientific explanations, I recognise some limitations. Insisting on 'transformation', for example, may not always serve feminist political ends. My insistence on transformability is for thinking about organisms. Genetic reductionism, however, also (somewhat paradoxically) permits discourses of transformation, by moving genes around, *within* the rhetoric of reductionism. Surgical transformation may not serve progressive interests, either; cosmetic or transsexual surgeries, for example, involve literally making the body over to achieve desired goals. But in neither is the material body thought of as having internal agency; rather, it is a fixed entity which is at odds with what is desired. Note that I am not saying here that thinking of agency will necessarily make the desired changes, simply that seeing bodies as reducible to interchangeable bits is part of the discourse of fixity.

Yet the fact that bodies are alterable within reductionist logic is not itself an argument against a political insistence on ideas of transformation and complexity. We need urgently to find ways of thinking about bodily processes (or about 'biology' more generally) that move away from simple reductionism, and that simultaneously allow us to theorise bodies lived *in* culture. In this chapter, I have sought to outline some alternative views, and to look at their advantages and limitations; altering bodies within reductionist logic is not the same as conceptualising organisms as *self* organising, and with their own integrity.

Nor are the practical and political consequences the same. As individual organisms lose integrity within reductionism (and increasingly so with the rise of genetic engineering and the creation of 'designer animals'), so they regain it within these alternative frameworks. For those of us concerned about both human rights and the rights or interests of non-human animals (see Birke 1994, 1997), retaining or regaining the *integrity* of the whole organism is vitally important.

Indeed, that is one of the reasons I am mistrustful of the cyborg as a liberatory figure. While Donna Haraway and others may see in the cyborg metaphor a potential for boundary disruptions (Haraway 1991a), I see in it the potential to damage through fragmentation. The damage lies precisely in the loss of organismic integrity. In the discourses of biology, it is genetic reductionism itself which is most 'postmodern' in its capacity to fragment. Yet, as I have argued above, there are constraints in the way organisms develop; they are simply not as diverse and fragmented as the story of genetic selection implies. It is that constrainedness which makes genetically manipulated organisms potentially disastrous (Newman 1995): and it is that constrainedness which would limit the possibilities for any (literal) cyborg. Even as a figure of the imagination, the cyborg allies itself to a vision of biology that I believe to be inimical to feminist interests and ideals.

The search for alternative models, for different (and better) stories to tell, lies at the heart of feminist theorising about biology. Feminists insist on more complex, nuanced, ways of interpreting biological processes. Partly, we do so because even empiricism allows different ways of interpreting evidence. Complex models better describe how things work: they also provide alternative narratives, in the postmodern sense, which challenge Enlightenment concepts of one truth (see Longino 1990; Hekman 1992).

A second reason for feminist struggles to rename nature through complexity and transformation is that we can thus challenge persistent dualisms. Seeing gender opposed to the bedrock of sex is one example. Others include the dualisms of organism/environment, human/

animal, bodily fixity/cultural lability, nature/nurture, and so on. As feminist critics often note, dualistic thought is deeply problematic – not least because it feeds dualisms of gender.

In opposing reductionism, and ensuing dualisms, feminists must insist on the uncertainty and indeterminacy of bodies (indeed, of all that we might characterise as 'biology'). And we must do so without relegating the internal body to an unquestioned realm of biomedicine. Yet we must also recognise that indeterminacy and transformability are not without limit. Bodies may constantly undergo interior change, but within apparent sameness. Perhaps nowhere is this more apparent than in the bodies of those with physical disabilities. Transformation may be the *modus operandi* of the body's interior, but it is unlikely to lead to sudden changes to physical disability.[64] And nor will thinking of bodies in terms of transformation alter the present cultural reproduction of disability.

While recent feminist work insists on cultural contingencies in describing bodies as marked, as signifiers of culture, it rarely goes beyond bodily surfaces as I have noted. Culture is inscribed *on* those surfaces. In doing so, we run the risk, as Elizabeth Grosz rightly recognises, of leaving the body's interior in the realm of biological fixity (Grosz 1994). Yet that risk rests on how we conceptualise biology itself. Bodies are good to think with only when we think of indeterminacy or transformation; otherwise, they disappear into an assumed and underlying fixity. Indeed, we might begin to think of our biology as offering potential rather than limitation, that it might be 'our biology that makes us free' as part of an engagement with our worlds rather than determiner of them (Rose 1997a, p. 309). 'Biology' need not always be the ultimate limitation.

8

Connections: the Body's World

The problem is
to connect, without hysteria, the pain
of any one's body with the pain of the body's world ...
(from 'Contradictions: Tracking Poems, part 18', Rich 1986)

My starting point for this book was the way in which 'the biological' slips out of feminist theorising, with its emphasis on the body's surface. So, my aim here was to begin to think about the inner body through the stories told about it in biomedicine. In doing so, I sought to explore their cultural location and their relationship to other narratives (the heart provides particularly clear contrasts between the language of biomedicine and other metaphors). I also sought to counteract the passivity of the biological body that seems to me to lurk behind many theoretical accounts, and to point towards less passive or less reductionist ways of conceptualising 'biology'.

Why does this matter? First, it matters because ignoring or playing down the biological, material body helps to perpetuate dualisms of mind versus body, with all their gendered connotations. Second, by relegating the biological to another realm we indirectly reinforce bio-logical determinism – a point to which I return below. It is because of that risk that I advocate a turn to other ways of thinking about biology, to move us towards 'better' stories, as it were. And third, a theory which posits biology as fixed/passive – even if implicitly – supports political practices which fail women. These are the themes that I want to pick up on in this chapter, sketching connections back to other work on ideas about the body and to the wider, political, implications of the biomedical stories I have sketched. In particular, my plea is for a view of the biological body that embraces transformation and change, and prioritises the body in relation – thus emphasising the connections of the 'body's world'.

In earlier chapters, I explored some of the stories science tells about our bodily insides. I focused on particular kinds of representation – in diagrams and through analogical models (such as 'systems') – and traced the ways in which these images or stories developed out of particular social and historical contexts. Those contexts have shaped the images that we now see as commonplace in textbooks, and which help to structure how most people in Western culture imagine the inside of our bodies to be. If military developments helped to shape ideas about how bodies work, then it is hardly surprising if, say, pictures of the 'fortress body' emerge from people's imaginations (see Martin 1994).

I have, of course, throughout this book, been concerned primarily with the stories science tells, with narratives and their social construction. So, I have written for example about the metaphor of control/regulation in the development of central ideas in physiology. In that sense, I have taken a social constructionist approach; it is important to recognise how scientific 'facts' *are* socially and culturally embedded. But these ideas, I insist, do have a relationship to biological materiality, to the messy body of blood and guts, bones and sinews – a point which often slips out of the pages of social constructionist analysis of science.[65] My reading of that biology is, to be sure, mediated through the wider culture – it can never be an innocent reading. But it is also a knowledge gained from getting my hands dirty in the materiality of the laboratory,[66] however much that practice is itself deeply cultural.

We make practical use, furthermore, of what we have come to understand about control in biology. Injecting insulin means that the levels of sugar in the blood of someone who is diabetic can be controlled, so making glucose available throughout the body. With appropriate drugs, we can now control the catastrophic events which signal shock and which would otherwise be fatal. And millions of women control their fertility by means of hormones to suppress the feedback that would otherwise result in ovulation. These are steps which intervene materially in the processes we have come to conceptualise through feedback systems.

In making that appeal to include biological materiality, I risk making too strong a distinction between science/nature and the metaphors scientists use. Cathryn Vasseleu (1991) points out that a strong distinction serves only to legitimate the scientific enterprise by separating out metaphor from the 'real stuff' of life. Science can thus carry on arrogating to itself responsibility for 'naming nature'. I do not, however, wish to lose 'the real stuff' altogether in mapping out how the body is understood in Western culture; whatever metaphors we use, the biological body carries on its biological processes. The way in which we represent – through language and diagrams – the workings of those processes is, however, a cultural product. Those representations affect not only how we come to understand the biological body, but also how we live it (Diprose 1994, p. 129).

Similarly, the forms of representation used to narrate 'how the body works' themselves affect how we live our bodies. We inherit a language of biomedicine that insists on reducing our bodies to constituent parts; how we live our bodies is intimately affected by the power of that language. Can anyone now, in the West, experience severe pain in the chest without imagining that specific organ, the heart? We have learned to compartmentalise our bodies, and thus our experience of living them.

The purpose of this concluding chapter is, then, briefly to revisit the major themes of earlier chapters – the narratives of space, control, and information – and to link these to wider questions of politics, and to the idea of 'living the body' as a biologically material one. In particular, I want to elaborate on the theme of my critique of the passive/fixed body, and to emphasise transformability and relationality.

UNLIMITED SPACE

The metaphor of space and the spatialisation of visual diagrams in science was the theme of Chapters 3 and 4. I noted there how increased abstraction from context has characterised the images used in scientific diagrams, and how that might structure our reading of those images. In particular, I dwelt on the space that the images contain, and on what that might subliminally convey. I remarked how I have sometimes encountered surprise on the part of students that internal organs are packed so tightly, for they had a sense of bodies as having inner spaces.[67] Because of that, I traced the ways in which representations of inner bodies perpetuate the 'space' metaphor.

The spatialisation of the body described by Foucault (1973), in relation to medical understandings of pathology, served to localise disease within the body.[68] Thus, doctors seek to locate disease within specific organs, even when symptoms occur in widely separated parts.

Disease becomes fixed into specific locations. As the inner body became mapped, its structures indeed became more localised. Yet paradoxically the ways in which those structures were represented helped to perpetuate the metaphor of space within the body (a metaphor which also resonates with many other ways of conceptualising bodily function, in both Western and non-Western cultural traditions[69]). The circulation of this metaphor within the wider culture, moreover, is facilitated by the science that would deny it. For while scientists might insist that bodies were packed full of organs, they continue to represent bodies as though they were full of space. Organs float in the void of two-dimensional diagrams.

One, perhaps beneficial, implication of thinking about inner bodies in terms of inner space is that it reminds us of the continuity of tissues. It thus works against the profound reductionism that insists on seeing everything within the body as separable or even fixed. Biologically, what happens in the body *is* continuous and interdependent; processes connect to one another, even if particular structures do not. On the other hand, there are implications of the space metaphor that are less desirable in terms of feminist politics.

My reading of anatomical diagrams in terms of space links to Elizabeth Grosz's analysis of gender and the body (Grosz 1994). There, she asks whether 'the female body has been constructed not only as a lack or absence but with more complexity, as a leaking, uncontrollable, seeping liquid; as formless flow; as viscosity, entrapping, secreting; [as] ... a formlessness that engulfs all form, a disorder that threatens order?'. Women may have many anatomical structures in common with men, she notes, but 'insofar as they are women, they are represented and live themselves as seepage, liquidity' (ibid., p. 203; see also Young 1990).

The fluidity to which Grosz alludes is (partly) seepage out from the body – from the breasts in lactation, from the genitals in sexual desire. To be female is to leak in excess.[70] That formlessness finds echo, I suggest, in the implicit formlessness of the medical diagrams. Space, of course, characterised diagrams of both female and male in the illustrations I examined. But the diagrams of male reproductive anatomy sometimes convey a solidity lacking in those of females, as I noted in Chapter 4. Moreover, as long as female anatomy is portrayed as secondary to that of males, then the replaceability of the uterus will be reinforced. Human anatomy is conveyed in elementary books and models as universal 'man'; woman is the add-on extra, the one with the removable uterus. It is precisely that role as the anatomical extra that reinforces the formlessness – or space – of the inside of women to which Grosz refers.

In that sense, the very abstraction so typical of the scientific diagram itself accentuates gendered difference. In the formlessness of diagrammatic space we can lose what human bodies have in common. Thus, most diagrams in elementary textbooks show an obviously female outline only if what is shown swimming around in the inner spaces is a female reproductive tract. Guts, lungs, hearts – these are more typically illustrated within a 'male' outline or no outline at all. In that sense, difference of gender becomes fixed and accentuated through the specialised medium of the anatomical diagram.

Furthermore, the abstraction allows us more easily to imagine human bodies as a collection of interchangeable – and removable – parts. This then encourages a *focus* on the organs that can be removed – the uterus in hysterectomies, for example. One consequence of this focus in terms of gender is a foregrounding of anatomical gender difference, at the expense of seeing what biologically we all have in common. It is precisely gender difference that emerges out of the pages of scientific textbook illustration. Commonality and overlap are played down in both text and diagram.

Concepts of inner space lend themselves to greater interventions, precisely because of the focus on interchangeable organs free-floating within. These can take the form of various surgical procedures (ranging from unnecessary hysterectomies to the increased interventions imposed by techniques such as *in-vitro* fertilisation). Visualisation technologies, too, are designed to probe even into the otherwise healthy body. As Margrit Shildrick notes, a great many 'healthy' changes to the bodies of women are brought under medical scrutiny, in childbirth for example. This 'speaks to a deep cultural unease with the embodiment of women ... [manifest in] the ever-present reality of potentially hostile external intervention into their body spaces, and into the space of their bodies.' (Shildrick 1997, p. 170).

Inner space is not, however, always easily localisable – as indeed many of our experiences of bodily sensation are not. Pain is one example; it cannot itself be visualised by any form of technology. While some pain may seem to originate in a specific place, much pain does not; we cannot seem to fix its spatial coordinates, still less to try to describe it to anyone else. It is precisely our inability to put words to the meaning of pain that renders it untraceable by biomedical practice (Scarry 1985).

It is just that untraceability and unspeakability that, argues Scarry in her analysis of the body in pain, make the infliction of pain in torture so possible. Scarry notes also the metaphoric extension of bodily space into the space of the torture room – space within becomes coterminous with space without. A house or room can be an extension of the body, protecting us and helping our bodily homeostasis by keeping us warm and dry (and links to the architectural metaphors noted in Chapter 3). But the meanings of that protective space can be dramatically turned around, so that the very room itself can become an instrument of the torturer: 'it is not just the space that happens to house the various instruments used for beating and burning and producing electric shock. It is itself literally converted into another weapon, into an agent of pain' through brutal use of doors, windows, furniture to inflict suffering (ibid., p. 40). Inner and outer space become blurred in the moment of pain.

More abstractly, the metaphor of space as a way of imagining the inner body thus resonates with space outside – not only of rooms but also of space beyond the body. Romanyshyn (1989) speculates about these connections in his exploration of technology as a history of the discovery of linear perspective. This development, he suggests, paved the way for the ideal of the detached spectator and the body as an observed specimen. It was not accidental, he suggests, that the anatomical corpse was emerging as an object of inquiry just at the point in history when 'with Copernicus, we transformed the earth into an object in space' (p. 147). This object of inquiry was also emerging, we might note, at the same time as the spatialisation/localisation of disease within the body described by Foucault (1973).

Through these linkages, the anatomical body becomes the body of the astronaut, which 'sums up the dreams of reincarnation and departure which have animated the discovery–invention of the abandoned body [of the corpse]' (Romanyshyn, ibid., p. 148). In this view, then, the spaces mapped out in anatomical illustrations, and imagined to exist within us, map also onto outer space.[71] We project each onto the other.

Moreover, inner space and organs connect to the space outside us, if not to outer galactic space, through global communications technologies. Not only are brains (minds) in communication through such means, but other organs can become connected; thus, a doctor on one continent can view the diseased organ of a patient in another part of the world. And we can all view the organs of the corpses which became the Visible Human Project.[72] But, again, these representations are of organs essentially sealed off from one another: connected, but not continuous. To pursue the metaphor of galactic space, it is as though we see only the planets and not the overall relationships between them within the galaxy.

Representations of the inside of the body in terms of (outer) space voyages make frequent appearances in film and television documentaries. This was made explicit in the movie *Inner Space*, where imagery of space travel accompanied the voyage of a miniaturised human along the capillaries and byways of a human body. Here, technology allows us to probe both into the microspaces of the inner body as it can probe the macrospaces of galactic space. Underlying both is the motif of exploration, probing the secrets of those spaces.

Among the spaces we may thus probe is the uterus itself. It is perhaps no accident that the foetus is so often portrayed in photographs as a spaceship-traveller, as Ros Petchesky (1987) described it. The foetus becomes separated from its context, the uterus on which it relies and in which it is embedded. Through that portrayal, the focus shifts from the pregnancy of the mother to the foetus as a separate entity. This is made possible by the use of visualisation technologies making visible the previously invisible; it is also made possible by a prior conception of the inner body as space and by the resonance of that idea with our fantasies about outer space. Once such photographs were produced, they reinforced the conception of space, as though the foetus floats free like the astronaut. The way was then paved for the development of concepts of foetal rights, based on assumptions of foetal independence, and intensification of campaigns to illegalise abortion.

In mapping out these ideas of space and the fragmentation of the internal body they imply, I am speculating on possible connections within our cultural imaginations. It is, of course, a very long way from the sparse diagrams of textbooks to the horrors which humans manage to inflict upon others. I certainly do not wish to imply direct lineage: yet the fragmentation of the body that we learn from biomedicine does seem to enable such horrors. Indeed, the fragmented body of biomedicine might be said to underwrite the modern torture room, not least through the literal presence of supervising doctors to monitor the limits to which the body can be pushed (see Scarry 1985, p. 42; British Medical Association 1992). In that collapse of outer space onto inner space occasioned by pain, the supervising doctor embodies the visual gaze brought into the body's spaces themselves.

There are, of course, many other ways in which I might have explored the possibilities of mapping ideas about the inner spaces of the body onto ideas about space outside the body or vice versa. To imagine either at the end of the twentieth century involves visualisation technologies, which bring images of the space 'out there' (even if that is an image of our inner space, such as an ultrasound scan, we still have to see it as an image outside ourselves).

Imagining our organs 'out there' in some ways reflects the Western habit of separating the body from the transcendent mind. We can 'leave the meat behind' in cyberspace, or we can view it at a comfortable distance through imaging technology. These are, it might be said, the consequences of one of the dreams of Western science (not to mention theology), involving visions of escape from the mundane body (Midgley 1992). In this fantasy, there are no limits: perhaps one day, 'life could evolve away from flesh and blood and become embodied in an interstellar black cloud ... or in a sentient computer' (Freeman Dyson, quoted in Midgley, ibid., p. 150).

If that is the dream of Western science and technology, it is a dangerous dream; it has led, Romanyshyn (1989) points out, to the potential destruction of our world through such technologies as the atomic bomb. He invites us to imagine a different dream, in order to imagine a different world. Accordingly, I want to dream of the *limits* of flesh and blood, of the bodily forms of the world's living organisms, which have constrained how life has evolved over millennia (Newman 1995); this is my dream if I want to escape nightmarish fantasies of limitlessness and their consequences. As I noted in the last chapter, life, organisms, are not as limitless as Dyson's fantasy, nor as limitless as would be implied by the fantasies of genetic engineers. It is those limits which provoke in me both fear (once genes are moved around and there are unforeseen consequences imposed by fleshly constraints, let alone the possibilities for suffering that might result) and hope that flesh itself will constrain the fantasies of escape.

LIMITS, CONTROL AND INFORMATION

Let me now return from fantasies of escaping into space to the rather more prosaic themes of control, regulation, and information, addressed in Chapter 5, and to consider some of their implications. Unlike dreams of limitless expansion into the space of the universe, physiological controls and regulations imply limits.

The *uncertainties* and *unpredictabilities* of biological processes are written out of the narrative of control, and what emerges is a story of 'biology' as a set of simple determinants or constraints. And it is precisely the assumption that an inwardly controlled biology might control us that so concerns critics. For most feminists, the idea that we are simply passively responding to controlling genes or organ systems, like puppets on strings, is anathema.

The passive body is not only to be found in determinist biology, however. Writing about recent postmodernist work on the body, Terry Eagleton notes that, 'for the new somatics, the body is where something – gazing, imprinting, regulating – is being done to you' (1996, p. 71). He goes on to argue against that passivity, noting that 'What is so special about the human body, then, is just its capacity to transform itself in the process of transforming the material bodies around it' (ibid., p. 72).

The passivity of the gaze to which Eagleton refers is further reinforced by the notion that the senses operate primarily as one-way traffic, to which I referred in Chapter 4. I want to note two implications of this view here: the first is that the emphasis on inward flow facilitates a view of the sensory systems as separate from each other, almost atomistic. And second, the implication of the one-way motif is that we are passive recipients of 'information'; as such, it militates against agency in just the way that Eagleton implied.

We speak of the 'five senses', thus implying their separateness (though we also know that for many people it is possible to associate colours with sounds, smells with visual images). That separation, combined with the growing reliance in biomedicine on the conception of the body as replaceable parts, contibutes to a perception that those who do not seem to have 'five senses' somehow lack something physically. 'Lacking hearing', for instance, seems to imply something missing, a bodily deficit, a gap in the inward march of information.

One of the tasks modern biomedicine takes upon itself is precisely the correction, often through technology, of such perceived bodily deficits – to 'make up for' the lack. The pursuit of this task takes for granted that there is a lack, that somehow to be fully social requires a particular configuration of information flow. So, for example, the development of technological devices aimed at 'enabling the deaf to hear' presuppose that it is desirable to convert 'the deaf' into hearing people, to 'correct' their 'lack'.

Examining the development of the technology of the cochlear implant, Blume (1997) notes media rhetoric which spoke of the 'bionic ear' that would, miraculously, enable even those who were born deaf to hear. But the scientific miracle was fiercely opposed by groups of Deaf people, who objected to being cast as sets of 'non-functioning ears', as though they were lacking in communication skills. As one such group noted, Deaf people do not need to learn to hear or to speak; they have their own language: 'We speak only with a gestural language, that is, our mother tongue, and it is marvelous to speak this language' (cited in Blume, ibid., p. 46; see also Wendell 1996, p. 75).

It is precisely the construction of the senses as both separate and unidirectional that enables this notion that we can 'lack' important information, thus 'disabling' us from proper social engagement. Just as disability rights groups often remind those of us who are (currently) non-disabled, it is society rather than the individual's body which creates disability. In the case of Deaf people and the medical development of cochlear implant technology, it is partly social acceptance of the predominant story of the senses as inward and passive; this story builds a deficit model of the person and her/his senses. Tales of deficits demean, denying agency.

The second consequence of the heritage of one-way information is that it enhances the concept of the passive body. The idea of the person as passive recipient of sensory information developed alongside the historical emergence of the techniques of surveillance charted in Foucault's work (1973). The panoptical gaze, the medical gaze, intensifies and controls; it gazes into us, just as sensory information (photons of light, sound waves, molecules of scent) seem passively to enter our brains. Moreover, the various forms of imaging technologies – ultrasound, CT scans, and so forth – serve to bring the inner control systems of the body to the gaze, thus exteriorising the hitherto unseen systems of control and regulation.

Yet we might write another story. Perception, for example, is not itself passive. On the contrary, we select from the arrays of information with which we are constantly bombarded; our brains do not simply receive but actively interpret, act upon incoming information.[73] I chose the example of the senses here to demonstrate how passivity – and hence fixity – becomes almost inevitable if information flow is one-way. In this example, the motif of passivity has been part of the development of scientific narratives, built into the history of optics and visual physiology (Keller and Grontkowski 1983). And it spills over into sociocultural accounts of the body, themselves rooted in the preeminence of the visual gaze. That heritage helps to perpetuate notions of the body as fixed, itself passively receiving – notions which underwrite feminist theory as much as they pepper the narratives of determinist biology.

As I noted in the last chapter, an emphasis on complex systems and the fluidities of information flow is one response to the dilemmas of determinism, control, and passivity. In principle these ideas put the biological body into context; we become part of, and integrated into, complex systems. The body seems thus to become less isolated, less passive.

But these ideas are not without problems themselves as I remarked earlier. In Chapter 7, I considered the problems of such views of complexity/information for thinking about the biology of organisms; here, I want to look at them in a wider context. Emily Martin (1994), in her analysis of cultural conceptions of immunity, notes the spread of ideas about the body as a complex system within popular culture. But, she points out, there is a paradox in that these ideas engender a kind of empowered powerlessness – you can feel more responsible for your own health if you see your own body as a microcosm nesting within other complex systems, but also powerless as those other systems become more controlling (ibid., p. 122). Control, moreover, may no longer be one's own; rather, the control of complex systems can suddenly switch elsewhere in the system, unpredictably. The result, argues Martin, is that people increasingly feel unable to influence 'the system'.

Barbara Duden (1993), writing about women's bodies and reproduction, finds similar rhetoric among the students she teaches, and suggests that perceiving the world in terms of systems is part of a sea change in conceptualisations of our bodies. Younger women, she argues, see the world differently from women of her (and my) generation. 'I am', says Duden, 'a member of the generation that never detached its own body from [the] pictures' on classroom walls of the inner parts of the body – skeleton, endocrine glands, and so on (ibid., p. 48) – producing an atomised understanding of our bodies.

This was little challenged, Duden believes, by what many of us saw in the 1970s as a radical act in relation to our bodies, a challenge to medical hegemony – self-examination using a speculum. Rather, the radical break came later, with a shift towards women 'seeing themselves in terms of feedback within a psychophysiological system ... [in various disciplines] students are taught to look for complex feedback and communication within a system. Their bodies are transistorised rather than transparent' (ibid., p. 49).

Duden argues that the new rhetoric facilitates a reconceptualisation of pregnancy. We no longer focus on 'quickening' of the foetus within the womb as the defining moment of pregnancy; instead, we have reified an abstraction called 'life', which we assume to begin at the point of fertilisation. The embryo/foetus, suggests Duden, is increasingly viewed as a cybernetic state, to which is attached the abstract – and rather meaningless – label 'life'. The advocacy of 'life' rather than women's bodies is facilitated by recent developments in biomedicine, she suggests. Women's bodies are no longer that which gives life, but 'life' is seen increasingly as belonging to genes. 'Life' can thus be manipulated, as scientists move DNA around between (apparently lifeless) organisms.

To speak of our bodies as cybernetic organisms (cyborgs) does help to disrupt once-dominant categories and so create new political spaces (Haraway 1991a). Yet it is also dangerous in that it seems to me to remain paradoxically rooted in denial of the biological/material. The emergence of the rhetoric to which Duden alludes is a product, I would argue, partly of those forms of representation within science that have constructed images of the body in terms of space. Alongside those images, came the highly stylised and abstract diagrams of feedback circuits which in turn yielded the rather formless notions of 'information flow' now becoming more predominant.

The danger for feminists lies in these connections, for it is in the transition from a fleshly uterus to a cybernetic 'life' that the politics of abortion emerge most clearly. The shift towards 'life' is clearly advantageous to those who would argue against abortion in any circumstances, and who would champion 'foetal rights' to life. Precisely because the potential to become human has been read onto DNA, concern may be expressed over what happens to fertilised eggs thrown away with the petri dishes in infertility clinics. That these zygotes cannot achieve their potential unless they become *implanted* in a living woman's uterus (and even that is no guarantee) becomes irrelevant. I emphasise 'implanted' to insist, moreover, that the zygote must become not only attached but integrated into the uterine wall for a successful pregnancy. It becomes a separable entity only as a result of the success of that earlier integration, which itself depends upon the body of a woman.

Duden's argument rests on her case for a 'sea-change' in attitudes within the wider culture towards the functioning of the body – its 'cyborgisation'. I have to say that I am less convinced of the sea-change; both Duden and, in slightly different contexts, Haraway (1991a, 1991b) have commented on what they see as this shift towards a kind of postmodern, cybernetic thinking. But the spread of these ideas has taken place against a background of greater atomisation (genetic engineering, say), and hence, control, as I argued in the previous chapter.

Nor have I found, in my teaching, that younger women have necessarily adopted a 'transistorised', or cyborgian, language to describe their own bodies, except in their more ironic moments. Whether you see a massive shift towards conceptualisation of the body in terms of fluidity and cyborgs probably depends upon perspective. What I would argue instead is that the shift described by Haraway and Duden *coexists* synergistically with explanatory frameworks based on control. Postmodern flexibility might be the order of the day in, for example, immunology or in genetic engineering, but not in some other areas of biology. The danger lies in how these different frameworks draw on one another; control can be reinforced through 'flexibility', as Emily Martin (1994) reminds us in her voyage between immunology and the discourses of corporate management.

It is the materiality of the body that makes the consequences of different theoretical interpretations important; how we come to view the inner workings of the body has profound implications for health, and for how we live our bodies. What, for example, are the consequences of describing the body in terms of postmodern boundary transgressions, of ever-fluid flows? As I have already implied, I have reservations about the ubiquity of this view of the body. I also have reservations about it in biological terms. For it seems to imply a kind of random chaos even when it is extrapolated to the material body; information (whatever that is) seems to flow through like a zephyr – tantalising but never touchable.

Yet order emerges out of such (apparent) chaos. What I find frustrating about the narrative of fluidity is that it seems to lose sight of the ability of the biological body to be self-organising and self-determining – to be transforming within itself. It also loses sight of the constraints of flesh (Newman 1995): organisms simply are not that fluid or flexible in the development of their body plans. The bodily transformations which I insist upon bringing into feminist understandings of biology are not, however, only the willed interventions of culture on the surface, but are also processes of change within the organism; they are processes, moreover, which are certainly not randomly chaotic, but generate order.

I remarked in the previous chapter how there is no well-developed theory of the organism in biology; on the contrary, it has been overwhelmed by the hegemony of molecular biology to the extent that genes become what we are within reductionist rhetoric. This approach loses sight of the complexity of the whole organism, and the ways in which structures are generated within it.

My reservations about such powerful reductionism are partly biological, as I noted earlier. I remain unconvinced that we can easily control what happens when genes are moved around between organisms, and am fearful of the consequences. Genetic engineering treats the organism as just another of those 'black boxes', similar to the ones beloved of physiological modellers. Once we have identified a gene 'for' a particular trait, the reasoning seems to go, we can move it to where we want it. This presupposes that the 'black box' functions in particular, predictable, ways. That is a bit like expecting your computer to function predictably every time you switch on – which, as many of us know, is a hazardous and foolish expectation.

Genes may be an important part of how our bodies function or how an organism develops into a particular form; but they are not the only part. The physiological or developmental processes of our bodies can themselves generate order that is not directly dependent upon genes; structures emerge in development that themselves help to generate and constrain new structures. And these generative processes are going on all the time, even in the adult body – though here they act to maintain form. Thus, moving genes around without adequate knowledge of how the gene will fare in its new location, in different contexts, is potentially dangerous.

The politics of genetic manipulation is clear enough; commercial companies producing genetically manipulated organisms publicise their belief in the safety of such manipulation, and the use of statutory controls. Environmental and health activists, by contrast, point to the uncertainties of genetic manipulation and the risks of damage to eco-systems or to public health if such organisms escape the controls.

There is, however, a second set of political consequences which are perhaps less obvious. A biology that loses sight of the whole organism is one that permits a view of organisms as sets of replaceable parts. We witness the domination of that view whenever we read about 'spare part' surgery, or about experiments creating, say, headless embryos which might be used as organ donors. It is a view that permits a conceptualisation of organisms, of bodies, as lacking intrinsic value except as a set of parts. At the time of writing, the British government has just given a cautious go-ahead to the further development of transgenic pigs as organ donors, despite opposition from many people concerned about the risks of disease transmission from pig to human. Now pigs are, of course, already treated as sets of parts in the production of meat. But here, in the case of xenotransplantation or bio-pharming (the creation of genetically engineered animals to produce pharmaceuticals), we are carrying that objectification a step further.

As someone concerned about the status of, and our relationship to, non-human animals (see Birke 1994, 1997), such moves are worrying enough for their increased objectification and commodification of the bodies and lives of animals. Relatedly, they also facilitate a view of human bodies as fragmentary, as not entities. There are, undeniably, medical benefits to transplantation; the recipient of a transplanted organ gains – potentially at least – a new lease of life. But the concept-ualisation of bodies *as* sets of parts lends itself to a loss of any sense of the body as an entity in and of itself, with an integrity that we might respect. To what extent is it any longer 'my' body when replacement parts, from whatever source, are inserted?

To be concerned about the loss of the organism as entity is a double-edged sword, as I argued in the previous chapter. I am well aware that the Western liberal tradition has generated an atomistic view of the self and of the body, in which bodies become separated from their world (Diprose 1994, p. 127). But to advocate fluidities and fracturings in recent social theory is to lose the organism/identity altogether.

My reasons for not wanting to lose the organism are twofold. First, I have argued that the extraordinarily powerful message of reductionism and the rise of molecular biology have ensured that we easily lose sight of ways in which bodies become structurally *organised* in the way they do. So, we lose much understanding about how biology works in the focus on molecules.

My second set of reasons are political. Fragmenting bodies discur-sively goes hand-in-hand with their literal dismemberment, whether well-intentioned (as in transplant surgery) or otherwise. If we concep-tualise our bodies as a set of bits, a kind of mobile set of building blocks, then we can imagine ways of coercing or persuading people that they must have a few of these bricks removed. One way of doing this (rightly or wrongly) is through the disciplinary power of medicine. Often, this will, indeed, save our lives; sometimes, it may be un-necessary, as is the case with some hysterectomies and sterilisations. And sometimes it can lead to human rights abuses, through the direct exploitation of people as sources of spare parts which can be bought or sold (British Medical Association 1992; Kimbrell 1993).

What we need to develop, I have argued, is an understanding of biological processes in terms of transformation/change. But this should not be a concept of transformation that celebrates a formless fluidity; on the contrary, what we need is alternatives, such as the less popular (and seldom popularised) stories of organisms and their biologies as self-organising. Organisation emerges in these stories from dynamic processes of change, and the organism (or the body) emerges as more than simply a sum of fragmented parts, retaining some integrity.

That is not to say that these stories are faultless. But some are better than others. We need to draw on these alternatives, I have argued, so that we do not lose the biological body altogether. But we need to do so in ways that do not lose sight of the organism's contexts, its *relationality*. For one of the dangers of focusing on 'the organism' – even on change within it – is that we can lose the context.

Part of that context is certainly the physical environment, which can influence and be influenced by, the processes of our inner bodies. But it is also the social relations in which we, and our bodily functions, are operating. These contexts of relationality are too easily forgotten when we come to think about biology and the body – which all too easily becomes a body hermetically sealed.

One reason for that separation, of course, is the long-standing separation of the disciplines of academic inquiry. There are few ways of thinking through in detail how processes within the body might engage with social relations (except, significantly, within the rhetoric of 'stress', the 'privileged pathology of communications breakdown' (Haraway 1991b), which can alter immunological responses).

One possibly useful route to this end is suggested by Diprose (1994), drawing on Merleau-Ponty's concept of corporeal schema (which links the body–self to the outer world, another kind of spatial mapping). This concept still prioritises the individual, subject to possible influences from outside; nonetheless, it does attempt to remove the body from its social isolation and to link it experientially to the outer world. Diprose suggests that in illness, the corporeal schema '"goes limp" ... [which] suggests that illness represents a breakdown in the structure of the self' (1994, p. 106). The control systems have thus failed to maintain (unremarked) homeostasis; they are not behaving normally.

Diprose emphasises that the concept of a corporeal schema is an expression of living in the world, so that it is at once an expression of living 'in' the biological body and a product of social engagement. On that view, medicine must deal not only with the breakdown of control circuits in illness, but also with 'the collapse of one's social expressiveness ... [thus giving] medical practice its ethical dimension' (ibid., p. 109).

As I have noted, the legacy of control systems terminology implies a body almost hermetically sealed from the outside. To be sure, physiology textbooks will explain how particular physical variables can affect the control systems; changes in the pressure or composition of the air we breathe can fundamentally alter characteristics of the blood and circulation for example. But the environment outside us does not much enter the equations, as though we are to an extent insulated from it. In particular, the *social* world we inhabit is not part of the story: in the framework of physiological models, it does not and cannot alter the outcome.

Indeed, Diprose (ibid., p. 124) points out, the potential effects of the physiological models and discourse on our engagement with the world as lived bodies remain unclear. As I have tried to show in this book, the forms of representation in biomedicine, as part of the culture we inhabit, have great potential to determine how we experience our bodies, in both health and disease. Biomedical discourses structure difference as deviance from a 'norm', thus building into the theory the foundation for discriminatory practices (Diprose, ibid., p. 127).

The idea of a corporeal schema does not, as Diprose outlines it, include those physiological processes as such. Given the rhetoric of biomedicine, the possibility of bodily changes that result from our engagement with the social world slips away. We know remarkably little – even within the framework of biomedicine itself – about how such social engagement might affect us bodily.[74] Take one example, from the area of biological research in which I have worked – hormones and their effects on brain and behaviour. There is an enormous literature on how this or that hormone affects this or that behaviour, in various species. But there is a much smaller literature documenting experiments which focus on how behaviour and social engagement might affect hormones or internal organs.[75] And if that is the case with laboratory animals, then how much more so for humans for whom 'social engagement' is so complex.

We might similarly draw on Ingold's (1990) idea of the human infant developing within a nexus of social relationships which create a 'field' (rather like the fields I noted in Chapter 7 that help to guide embryonic development). This field then helps to shape perception and behaviour. While Ingold's point referred to a behavioural field, the idea of such fields could be extended to the physiological processes of the biological body as well:[76] it is just that relationality – between organisms/selves and between the long-separated mind and body – that I want to emphasise, and which drops out of the individualist stance of modern biomedicine.

Nonetheless, I believe that we need to begin to develop a view of the body which attempts to put biological processes into their social context, yet which retains organismic agency and integrity. That is a difficult intellectual task, but one that needs to be undertaken if we are to move beyond the perennial problems of such dualisms as mind versus body.

In her work on feminism and the body, Elizabeth Grosz (1994) employs the motif of a Möbius strip – a continuous figure-eight strip joined so that outside becomes inside. This image, she suggests, symbolises how we might think of the relationship between mind and body; they are coterminous, not separable. We might similarly use the Möbius strip as a symbol of inner and outer body – not in the sense of the mind as 'inner' in the psychoanalytic way in which Grosz refers to it, but as the inner physiological body.

We might think of the skin as providing one example of such a link. It is both inner and outer. It is an organ in its own right, yet it is part of our exteriority. Yet it is also typically described as an impermeable *layer*. This story unfolds in popularised biomedical narratives – your skin is there to protect you against a hostile world. It is also a layer in much feminist theory, where it is simply an inscribable surface. Within the wider culture, the skin is deeply saturated with meaning – difference marked by the colour of the skin, or by ways in which we clothe or mark it. Yet it is also in the attribution of so much meaning to surface qualities that the skin becomes fixed into a layer.

Even to speak of 'the skin' as a unitary entity is to fix it, however changeable it may seem on the surface. Yet as a bodily organ it is not fixed; rather, it is both a protective layer, a surface, and a continuously mobile and changing internal organ. There is no clear separation of inner and outer, and the structure – and meanings – of the layers of the skin changes the deeper we probe it. In that sense, it can serve as a kind of Möbius strip.

We need, I believe, an understanding of the biological body that links inner and outer, rather than presupposing a singularity to the body. We need, moreover, a biology that is not determinist, nor is seen as foundational or presocial. But nor do we need an understanding that simply dissolves boundaries in endless flows of information. Our biological bodies are certainly not hermetically sealed; rather, they are in constant engagement with the 'body's world'. But their very material structure itself – the flesh – helps to structure, even constrain, how that engagement takes place. In that sense, information may indeed flow through the body, but not formlessly. In the case of electrical currents in nerve cells, for example, the nerves function as conduits and electrical charges flow directionally. Order emerges out of this potential chaos, as cell influences cell and specific patterns of influence form.

Given my background and experience as a biologist, it was perhaps inevitable that I should react vigorously to what seems to be a constant denial of 'the biological' in much social and feminist theory. In response, I have always tried – as I do here – to bring the biological back into feminism, while trying to avoid the pitfalls of determinism.[77] That is no easy task, and I am constantly and painfully aware of my own fracturing in the process. At one moment, I am exploring the social construction of gender; at the next, I must switch to writing about the very science I learned and practiced. I am both inside and outside the house of science (a kind of living Möbius strip!), with all the advantages and disadvantages that positioning brings. It is difficult indeed to learn to read 'against the grain' of what I was taught, to pick up on how I had learned to read the scientific stories and diagrams in particular ways and (equally importantly) what that process had done to me. At times in this book, no doubt, I have been less critical of the objectivist assumptions of that science than I might have been. I might, for instance, be accused of overoptimistic naivety in my pleas to seek 'better' biological stories.

Yet theories which deny the biological serve us ill, not least because it is through the biological body that we live in and engage with this world at all. But also, and significantly, our failure to engage adequately with biology (except to criticise it for determinism) fails those people (and non-humans) who are most readily defined by it, and who may suffer because of it.

Just as those of who are white are never called upon to situate ourselves in academic work *as* white (see Simmons 1997), so theorists of the body are never called upon to situate themselves as embodied. Yet it is precisely that erasure that makes the issue of the biological body a political one. 'Biology' is undoubtedly often made an excuse for discriminatory practices, and exclusionary discourses; it has been the ground for racism and sexism, for exclusions based on age or disability. Feminism must certainly reject that kind of 'biology'. Yet rejecting biological processes altogether by ignoring or omitting the biological body does not help; indeed, it merely serves indirectly to reinforce biological determinism.

Moreover, it serves to marginalise the embodied experience of those whose voices are not heard in science. Part of the critique offered by feminist science studies has centred on the ways in which women (and others) are excluded from science. That exclusion, Sandra Harding reminds us, ensures that science represents only the interests of a small élite, so generating a partial kind of knowledge (Harding 1991). To her list of groups of people who might add other partial perspectives and so make a fuller and better science, we might add the experiences of different embodiments – of bodies mutilated through torture or starving through global injustice, for example. What other knowledges about how the inner body works might thus be created? What would happen to our models of tightly controlled systems and mechanical pumps if other narratives of the body's insides had been admitted?

And by admitted I mean acknowledged fully as knowledge claims, not studied as quaint examples of 'folk beliefs' as some studies in medical sociology or anthropology do. Feminist critics, among others, have often bemoaned the lack of democracy in science – not only in terms of who gets to do it, but also in terms of the voices who contribute to it.[78] How we come to understand the body's inside is a complex product of the society we live in; in the cultures of Western industrialised societies, that understanding takes particular forms. We draw that picture partly from biomedicine; but less obviously the picture depends on wider cultural images and metaphors. In that sense, all of us contribute to the science; those multiple, and often discordant, voices should be heeded more fully in the creation of science.

Biomedical discourse is not, as it is sometimes portrayed, just another master narrative, despite its power. It contains within itself all kinds of contradictory strands and fragmentary positionings. It is a motley collection of narratives, building its tales on a motley collection of human and animal bodies; it is not a unitary story. Among those are tales that are less reductionist and which might serve feminist political ends rather better – and which can be more truthful stories of 'how bodies work'. Until such a time when the dream of a more inclusive science is realised, and when other stories of bodily insides have greater currency, insisting on other stories may be all that we can do. And by tracing the connections, we might finally begin to notice the inner body and to discover the body's world.

Notes

1. Clearly, there are many other uses of the word 'science', including the social sciences. Here, I am using it to indicate 'science' in the narrow sense in which it is often colloquially used (for example, in school curricula). I am well aware that other knowledges of the natural world may have equal, or greater, usefulness in describing that world; modern Western science is only one such story (see Barr and Birke 1998).
2. Among other things, it taught me a profound separation from the natural world that I love. That separation comes from the pursuit of objectivity (see Keller 1985), and from the practices encouraged by objectivity. Among other things, those include the belief that we are justified in using animals and their bodies as objects of inquiry (see Birke 1994, 1997).
3. Reductionism and objectivity are part of the rhetoric of science; neither are achievable in practice. On the contrary, scientists may speak in the laboratory of the contexts in which, say, a hormone or gene is operating, yet omit that from the written reports.
4. Although there is still controversy about the number of genetic mutations involved in such diseases, there is also considerable variation in severity of these diseases, which means that it is difficult to link gene to disease in any simple way.
5. That is, geneticists do sometimes claim to recognise the need for context. Yet, at the same time, many scientific reports – and even more popular reports – of developments in genetics speak directly or indirectly of genes 'for' particular traits. And, of course, some scientists use that language very directly.
6. These have been themes in my own scientific research (see, for example, Birke 1989).
7. These include Ayurvedic medicine, acupuncture, herbalism and many more. The approaches are often quite different, but my point here is to

177

emphasise how people increasingly turn to 'alternative medicines' as part of a resistance to allopathic medicine.

8. This was something that struck me and my colleague Jean Barr during research into women's perceptions of science (see Barr and Birke 1998). Discussing 'what happens in childbirth', all the women to whom we spoke mixed brief phrases from the 'scientific' story with more experiential narrative.

9. Although within the colonial medicine of the nineteenth century, there was a clear distinction drawn between the physiologies of Europeans and indigenous peoples, to create differences between 'black' and 'white' physiologies (see Anderson 1992; Gilman 1992; Arnold 1993).

10. In particular, postmodernism calls into question dominant dualisms and 'grand narratives', including that of science. As such, it helps to break down the apparent certainties bequeathed to us from the Enlightenment pursuit of rationality (see Shildrick 1997, pp. 5–6).

11. Darwin puzzled over inheritance, suggesting hypothetical 'factors' that were passed down from parent to offspring. Once Mendel's work with the genetics of peas was (arguably re)discovered around 1900, however, other experiments with plant and animal breeding quickly followed. The concept of 'the gene' developed soon after, although it has remained a highly contested concept ever since.

12. Oestrogens, progestins and androgens are the generic groups of steroid hormones. Oestrogens (e.g. oestradiol), and progestins (e.g. progesterone) predominate in secretions from ovaries; androgens (e.g. testosterone) predominate in testes. The secretion of all these hormones is controlled by other hormones, produced by the pituitary gland at the base of the brain.

13. In my lab, we worked on a hormone called medroxyprogesterone – used, at the time, as an injectable contraceptive. If infant rats consumed this stuff in their mothers' milk, the females developed 'masculinised' external genitals (see Birke et al. 1984).

14. One, earlier, version of the hypothesis suggested that what was crucial to 'make a male' was a good dose of testosterone. This was required to shift foetal development away from the basic, androgynous, body plan. According to this version, no such push is required to make females. This story has acquired the status of myth in feminist writing, where the basic body plan is seen as female. Even if we accept some of the organisation hypothesis, there is evidence that hormones are also needed to push development in a 'female' direction. Because of that, it seems more accurate to describe the basic body plan as androgynous rather than female.

15. This is not to say that all cyclicity is abolished, just that a particular cycle (the 28-day one in humans) disappears.

16. There are, of course, several other possible forms of representation of the physiological body, including three-dimensional models (see Jordanova 1991) and photographs (particularly in higher level medical texts, where

photography is used to illustrate pathologies or diseases). I have chosen to concentrate here on two-dimensional diagrams because they are the form of representation that is most common in elementary texts – and thus have the widest circulation.

17. Deirdre Janson-Smith, personal communication. The exhibition was 'Science For Life', focusing on the human body and produced for the Wellcome Trust in London.

18. Writing in the nineteenth century, the French scientist Cuvier used the word 'foyer' to speak about processes of assimilation of materials into the animal body – a word clearly conveying architectural space. 'Foyer' is both a place and a process in his writing, referring both to the hearth and to the processes of combustion. Living bodies, he suggests, are foyers 'into which dead substances are successively brought' (Cuvier, Lecons 1800–1805; quoted in Figlio 1976, p. 39).

19. I have encountered this belief several times in the course of running women's health workshops. How widespread it is I do not know, but it is certainly still extant.

20. Yet the power of mechanical metaphors was not created *de novo* in the Industrial Revolution. Pouchelle's analysis of the thirteenth century writings of doctor–surgeon Henri de Mondeville draws out the rich complexity of the metaphors he used to describe the inner body. Society and the body was one reciprocal source of metaphor; another was the imagery of buildings. The latter was in turn related to persistent images of nested, or embedded structures. Intriguingly, she notes that Mondeville used architectural metaphors repeatedly – but only with reference to the healthy body. By contrast, the sick body was metaphorically associated with things 'of nature', not structures built by people. Disease might thus be associated with animals or plants, or even minerals. Cancer, the crab, is an association with us still.

21. The desire to use pictures to persuade is not new in anatomy. Medieval anatomists certainly preferred pictures, according to Pouchelle (1990, p. 25): 'there are few things within these two fields of knowledge [anatomy and surgery] that can be represented in words' – so the anatomist and surgeon Mondeville believed.

22. I concentrate here on two-dimensional representations, because they form the basis of textbook illustrations and diagrams. Ludmilla Jordanova (1991) has documented the use of wax models as teaching aids for anatomy, and the way in which these models incorporated – literally – prevailing beliefs about gender. Also see Petherbridge and Jordanova (1997) for analysis of the associations between art and anatomy.

23. Dickinson (1892). The illustration was included in a paper presented by Lynnette Leidy (1994), 'Social Roles and Uterine Position: nineteenth century therapeutics for prolapse', given at the Five College Women's Studies Research Center, Fall 1994.

24. To supply the tissues with oxygen, the oxygen molecule must be able to detach itself from the carrier molecule, haemoglobin. This occurs under particular chemical conditions in the blood, and is represented graphically by what is called a dissociation curve.

25. The simplest kinds of vertebrates are usually referred to as 'primitive'. Given that animals somewhat like these were probably ancestral to all vertebrate types the use of this structure to illustrate typical vertebrates may be appropriate. It is the highly stylised image that I emphasise here.

26. According to a report in *New Scientist*, published at the time of writing this book, 'A new anatomical study shows there is more to the clitoris than anyone ever thought' (Williamson and Nowak 1998). The authors are both women. I rather suspect, however, that most women had 'ever thought' that there was more to the clitoris than scientists admitted.

27. The fragmentation of reflex movement is suggested, for example, by the photographic studies of human and non-human locomotion by Edward Muybridge at the end of the nineteenth century. These present us with what has been called a confrontation between bodies and machines, emblematised by the juxtaposition of moving bodies and gridded backgrounds in Muybridge's photography. See Seltzer (1992, p. 160) for discussion of this point.

28. The significance of the gaze in recent theorising derives in part from the predominance of vision in modern Western culture. That this pre-eminence of vision over other senses is gendered has often been noted (Keller and Grontkowski 1983; Irigaray 1985). Keller and Grontkowski explored the historical splitting between 'the mind's eye' central to the act of knowing, and the 'body's eye', or how light is picked up by the retina. They begin to chart visual metaphors through the work of Plato, who, they note, argued that not only do the eyes perceive images, but they also emit rays which come into contact with the perceived object.

    This understanding was central to Plato's metaphysics. But with the rise of modern science, that dual role of the eye began to disappear. 'With modern theories of optics', note Keller and Grontkowski, 'the eye becomes a passive lens, no longer thought to be emitting its own stream, and the transcendent coupling between inside and outside which Plato had imagined to occur was gone' (ibid., p. 213). Descartes, particularly, severed the link; for him, the (human) soul was separate from the body, including the eyes. Light and vision became more technical, and the eye became more and more a mechanical device, so that 'the active knower is forced ever more sharply out of the bodily realm ... Having made the *eye* purely passive, all intellectual activity is reserved to the "I", which, however, is radically separate from the body which houses it' (ibid., p. 215; emphasis in original).

    That radical separation leads to the notion that the knower can stand apart from *and gaze at* the known. The subjectivity of the knower simply does not enter the act of seeing. Nor does the body itself, for it becomes merely a passive receptacle for sensory impressions. Constructing the senses as passive and one-way, combined with the reliance on one specific sense, further supports the notion of the body as internal space; it is space to be filled – with information, with the (biomedical) gaze. And the gender of the visual gaze further reinforces the subject status (and genderisation) of the body's inner spaces.

29. These images are very similar to the drawings of sections of women's torsos produced by the eighteenth century gynaecologist William Hunter (whose collections of models and specimens were contributed to the Royal College of Surgeons in London). Some examples of these illustrations of torsos – legless and headless – can be found in the iconographic collection of the Wellcome Trust (History of Medicine Library). Also see Petherbridge and Jordanova (1997).

30. In the exhibition, 'Science for Life', developed by the Wellcome Trust in London, visitors could walk through a model cell as though 'walking through' the endoplasmic reticulum. While undoubtedly a good teaching device, this also creates an image of the cell as so much empty space.

31. This is illustrated by several photographs from the nineteenth century in the Iconographic Collection of the Wellcome Library for the History of Medicine, London (10713–10716).

32. In her poem, 'Heart test with an echo chamber', Margaret Atwood (1992, p. 118) describes watching an ultrasound screen, noting the cultural significance of the image:

> This is the heart as television,
> a softcore addiction
> of the afternoon. The heart
> as entertainment, out of date
> in black and white.

33. Just as she would have to learn how to read the 'natural history' photography in another book; photographs are themselves constructed images, conveying particular representations of 'nature' (see Myers 1990a).

34. Some changes are, of course, undesirable. The increased risk of bone fractures resulting from osteoporosis is well known, and it seems that declining oestrogen levels do affect it. But there are other factors that are much less often discussed, such as levels of exercise. For further discussion of osteoporosis and the menopause, see Klinge (1997).

35. Cannon had earlier written about what he saw as 'medical evidence that humans were physiologically programmed for fighting' (Crook 1998, p. 277). Possibly he believed that social controls were necessary to hold in check humankind's combative instincts. For further discussion of the heritage of these ideas of biological causes of human aggression, see Crook (ibid.).

36. Warwick Anderson (1992) notes how the idea that the body could regulate itself was part of a shift in the practices of colonial medicine. Describing the experience of European doctors working in the Philippines, he notes how the prevailing belief at the end of the nineteenth century was that constitutions were in harmony with the ancestral environment; so, attempts to acclimatise Europeans to tropical climates were likely to lead to degeneration and death.

37. Critics objected partly to the militaristic language of examples used by sociobiologists; the latter, in turn, complained that the examples were 'only' hypothetical. Militaristic association was, however, fundamental to the thinking behind such games, long before they found their way into modern sociobiology. For feminist criticism of sociobiology, see Bleier (1984).

38. Linking electricity conceptually to the body invokes images of electrical powering *of* the body – an image used by Mary Shelley as Dr. Franken-stein channels lightning into the inert body of the monster. At the end of the nineteenth century, the notion of electrical invigoration of the body gained popular appeal; it was also, suggests Carolyn Marvin in her history of electrical communication technologies, a notion redolent of gender. 'Virility', she argues, 'long associated with terms like *force* and *energy*, *strength* and *vigor*, which also described electrical properties, was an area ripe for electrical theorising and therapeutic promise' (Marvin 1992, p. 131). Perhaps not surprisingly therefore, electrical devices were advertised which, advertisers claimed, could augment sexuality (for men) or control it (for women).

39. Specifically, the inside of the cell is normally negative with respect to the outside. During the action potential, the inside transiently becomes more positive.

40. Scientists are now trying to clone the gene(s) (DNA) that they believe to be involved. There is, Trumpler notes, an irony in identifying the DNA sequence for cloning. 'This string of 7,000 A's, G's, C's, and T's [the bases that form the language of DNA] is, for a molecular biologist, the fundamental representation of the sodium channel. Yet for all but the most practiced eye, there is no way to extract much information about the structure or function of the sodium channel from these letters – ironically, since in the conception of biologists, the genetic code is information in its purest form' (Trumpler 1997, p. 77).

41. Graphical representation also contributes to this. Graphs represent data obtained from groups or populations; the line of the graph thus represents the mean value obtained, while the spread of different values in the data (called variance) is (usually) represented by 'error bars' – which appear as little vertical lines above and below each point on the graph. These representations of spread in the population do, however, tend to disappear as graphs as simplified for textbooks – with the result that the mean values can be read as 'typical'.

42. Barnard (1996): quoted in BBC television documentary, *A Knife to the Heart*, Part I: 30 April.

43. Both heart and brain might be thought of as the centre of our selves. Yet, curiously, there are far fewer metaphors associated with the brain. Commenting on this disparity, Scott Manning Stevens (1997) notes that 'The brain, by comparison [to the heart], presents us with a different problem for the representation of the self altogether. It lacks a stylised visual symbol such as the heart. We know what a heart-shaped object would look like but not a brain-shaped one' (p. 275).

44. Illustrations from the iconographic collection, Wellcome Centre for Medical Science, London (Collection Catalogue 10713–10716); not surprisingly, the women are white, bringing to mind Richard Dyer's analyses of the construction of 'whiteness' through lighting in visual images (Dyer 1997). The captions reiterate the themes: 'The dissection of a beautiful young woman ... to determine the ideal female proportions' (10713); 'An anatomist meditates on the corpse of a beautiful young woman laid out on a table next to his desk' (10715); 'An elderly anatomist contemplates the heart that he has excised from the corpse of a beautiful young woman' (10716).

45. William Harvey was the sixteenth century physician to whom is attributed the idea that the blood circulates, rather than ebbing and flowing. From the notion of circulation to the idea that it might need a pump is a small step.

46. These are babies in whom the central nervous system has not developed normally. As a result, the brain is largely missing above the brainstem at the top of the spinal cord. Such babies are usually born 'alive' in the sense that they are breathing and have a beating heart, because it is the brainstem, not the higher brain, that coordinates these activities. They cannot, however, live beyond a few days, and some doctors and ethicists have argued that they should not, therefore, be classed as fully human or even fully alive. Here, it is the possession of a functional brain that becomes the definition of life. Following this argument means that they could be used as donors for organ transplants (see Lamb 1990).

47. The heart's pacemaker is a group of electrically excitable cells, with an intrinsic rhythm. Their electrical signals cause the spread of a wave of electrical changes across the heart muscle, which in turn cause the heart to contract.

48. The heart's electricity was first noted in the mid-nineteenth century; by the century's end, a Dutch physiologist, Willem Einthoven, had devised a machine capable of recording electrical changes from the heart from outside the body. The advent of such a machine, along with the X-ray, finally allowed doctors to peer at the living heart.

49. Press Release; 22 March 1995, Georgia Institute for Technology webpage.

50. Which I found at: http://www.webcom.com/hrtmath/IHM/Women Empower.html.

51. Perhaps this is part of a wider cultural emphasis on 'makeover' – of our living spaces, of our bodies – that seems to be a theme of modern Western culture: we are not content simply to live in rooms, but must engage in 'making them over'. Thanks to Maureen McNeil for making this point.

52. Turner also refers to internal *re*straints, that is, the 'control of desire, passion, and need in the interests of social organisation and stability' (Turner 1992, p. 58). Here, he draws on psychoanalytic theories; this is a rather different sense of both 'internal' and of 'restraint' than the sense in which he refers to biological constraints. The interiority of psychoanalysis may, or may not, map onto the interiority of the anatomical body; the point I want to emphasise, however, is that both, in his writing, afford *con*straint.

53. For reflections on the loss of respect this move entails, see the various chapters in Birke and Hubbard (1995). See especially the chapters by Anne Fausto-Sterling and Ruth Hubbard in that volume.

54. My somewhat jaundiced view of how her work is read is based partly on responses to Haraway in the academic literature, and partly on responses by many of our students in women's studies. Haraway herself also regrets the way that her fascination for, and focus on, science seems often to be overlooked by commentators on her work (personal conversation, Healdsburg, California; September 1996).

55. Myalgic encephalopathy, also known as Chronic Fatigue Syndrome.

56. Though there are exceptions. Shildrick (1997), for example, attempts to write about the body in disease and in relation to bioethics, without losing her commitment to what she sees as the advantages of postmodern thought in disrupting dualisms.

57. Huntington's disease affects the nervous system, causing loss of function and eventual senility.

58. I am not intending to imply purposiveness here. Whether this form of active engagement bears any relationship to the purposive self and to intent – and if so, how – are questions I cannot begin to address here. They do, however, have implications for debate about abortion, and hence for feminism.

59. Some of these ideas draw on the sciences of chaos and complexity (for example, Kauffman 1995). Kauffman argues from the premise that the extraordinary degree of order in the universe is simply too improbable to have arisen by random chance (as would be required by Darwin's notion of natural selection); our existence, he argues, is so massively unlikely on the basis of random chance that we are immensely lucky to be here at all.

Kauffman's focus is on self-organisation; even non-living systems can become self-organising, such that a new order emerges out of what had previously seemed chaotic interactions. Mixing certain organic and inorganic chemicals can, for instance, generate clearly defined patterns despite the (assumed) random movements of the constituent molecules.

And in Chapter 6, I noted the relationship between order and chaos in the way the heart works. Emergent order is a crucial concept here. First, it is crucial because it implies that there are phenomena that cannot be explained in terms of constituent parts; the movements of molecules of those chemical reactions cannot explain the ordered patterns that result. Thus, it flies in the face of reductionism. Second, Kauffman suggests that it is these newly emergent phenomena that can be subject to selection, rather than single genes.

Chaos and complexity do provide alternative frameworks, but are not themselves without problems (see Martin 1994). I do not wish to over-emphasise them here, but merely to note how complexity theory contributes to the ideas I am discussing.

60. By cooperative, Goodwin means that complex organisms arose in evolution as a result of cooperation between simple organisms, such as bacteria (see the work of Lynn Margulis: for example, Margulis *et al.* 1996). Cooperation among parts of a system can contribute to new levels or order – Goodwin uses the example of cooperation among ants in a colony which creates higher level order in terms of social organisation. This, he notes, is in marked contrast to the prevalent Darwinian rhetoric of organisms as competitive machine.

61. Organismic integrity and uniqueness are, by contrast, promoted by the view of organisms as self-organising. As Stuart Newman has pointed out, the 'dominant genes' story of how organisms develop leads inevitably to the idea of transgression of biological boundaries, as it generates a plethora of possibilities through natural selection. But in the alternative view that he espouses, most diversification of biological form arose early in evolution. What happened subsequently was a *consolidation* of those forms despite genetic mutations. 'According to this view', he argues, 'the intensification of uniqueness, rather than the open-ended production of overt difference, may thus be the hallmark of organismal evolution once it has left its early, "physical" state. This view implies, furthermore, that mixing and matching the biochemical capabilities of modern organisms by transgenic manipulations could be profoundly disruptive of species and individual identity and integrity in a fashion different from anything encountered during evolution' (1995, p. 220).

62. Relatedly, Lisa Weasel (1997) draws on feminist work on psychological development which emphasises the self-in-relation, to develop her own account of cellular biology. Thus, she emphasises the 'cell in relation', in its cellular context, to shift the focus away from the prevailing reductionism of most molecular biology.

Again, the concept of agency I use here is not intended to imply consciousness; rather, it is an emergent property of the way cells and tissues organise themselves. In that sense, such 'agency' could be found in wide range of organisms, even in the amoeboid slime molds (see Goodwin 1994).

63. She also suggests that much modern writing by male scientists is built on dreams of escape from the body, coupled with power fantasies (see Midgley 1992; see also Romanyshyn 1989). To speak of virtual reality as 'escaping the meat' is one example.

64. Except through the intervention of technologies such as prostheses which, of course, create a kind of cyborg figure.

65. It is, moreover, the very recalcitrance of nature that makes it 'messy', in the sense of being at times unpredictable; scientists may construct a language of certainty, while the laboratory animal, cells, DNA or whatever does whatever it pleases (see Knorr-Cetina 1983).

66. Again, I would emphasise that that knowledge was gained at the expense of animal lives. The nineteenth century experimental physiologist (and apologist for vivisection), Claude Bernard, referred to the laboratory as a 'ghastly kitchen' in which animal suffering could be ignored. Reflecting on the heritage of that research, Hilary Rose (1995) reminds us that: 'Debates about reinventing biology and the increasingly costly (and sometimes ghastly) kitchen of research are inseparable from debates about the nature of our culture and society' (p. 18).

67. Notably students having little background in science. Science students are more likely to have had experience of dissection or of anatomical models. Whether the use of computer or video simulations of dissection (a welcome move in terms of animal use) makes any difference to student perceptions remains to be seen.

68. Foucault's (1973) use of 'spatialisation' focuses on the *location* of disease into specific organs. Note that this is not the same as (although related to) my emphasis on empty space as a significant metaphor.

69. Many non-Western systems of medicine, for example, are rooted in understandings of the body in terms of energy flow rather than separation of organs. These more 'holistic' approaches include, for example, the bases of acupuncture and of Ayurvedic medicine.

70. Male ejaculation, she suggests, is interpreted differently; it is seen as a cause (of fertilisation) and through that becomes solidity – see Grosz (1994, p. 199).

71. The 'mapping' metaphor is a pervasive one. It was used by early anatomists (Petherbridge and Jordanova 1997), and is made clear by reference to anatomical *atlases* (a phrase still in use today). It resonates, too, with the notion of mapping the genome, providing a genetic atlas (see Haraway 1997). Mapping the body of anatomy thus merges with mapping the genetic body.

72. I am grateful to Mike Michael for suggesting this connection to global communications technologies.

73. An important idea here is that of efference copy. That is, that as the brain sends a message to, say, the eyes to move, it also sends a 'copy' of the message to the parts of the brain that receive visual input. So, the brain

can differentiate between visual input that occurs because of active eye movement (reading a line for instance) and that occurring more passively (the scenery apparently moving when we travel for instance). This implies active process (albeit in the language of cybernetics!). The history of this idea is outlined in Grüsser (1994).

74. And, of course, there is the problem of how to speak across the language divide. I am not sure how readily the notion of a conceptual schema which can 'go limp' could be incorporated into the language of biomedicine.

75. That the environment might affect internal organs is suggested by experiments with young rats. Infant rats exposed to more complex environments (with more for the animals to do and learn from) showed more growth of some internal organs than was the case for rats living in relatively impoverished conditions (see Black *et al.* 1989).

76. I noted earlier how, in animal studies, gender differences are very commonly attributed to hormones, ignoring possible effects of maternal behaviour (see Birke 1989). We might also interpret these effects in terms of fields of the kind invoked by Ingold: the mother, her social world, her diet, her pups, their hormones – all interact to create a specific field in which those infants learn to engage with their world.

77. I cannot, unfortunately, develop on my own the kind of 'new biology' for which I plead. It would take the efforts of many of us. My hope is, however, that more engagement of social theorists *with* biology might help to encourage the kinds of questions and hypotheses that could eventually contribute towards new frameworks of thinking in biology. Perhaps what I seek is a cyborg biology, a hybrid form.

78. A point I explore in more detail in Barr and Birke (1998).

References

Adams, C. (1994) *Neither Man nor Beast: Feminism and the Defence of Animals*, Continuum, New York.

Anderson, W. (1992) 'Where every prospect pleases and only man is vile: Laboratory medicine as colonial discourse', *Critical Inquiry*, 18, 506–29.

Arnold, D. (1993) *Colonizing the Body: State Medicine and Epidemic Disease in Nineteenth Century India*, University of California Press, Berkeley.

Atwood, M. (1992) *Poems 1976–1986*, Virago, London.

Atwood, M. (1994) 'The female body', in P. Foster (ed.), *Minding the Body: Women Writers on Body and Soul*, Anchor Press, New York.

Balsamo, A. (1996) *Technologies of the Gendered Body: Reading Cyborg Women*, Duke University Press, Durham, North Carolina.

Barr, J. and Birke, L. (1998) *Common Science? Women, Science and Knowledge*, Indiana University Press, Bloomington.

Basar, E. (1976) *Biophysical and Physiological Systems Analysis*, Addison-Wesley, Reading, Massachusetts.

Basch, S. H. (1973) 'The intrapsychic integration of a new organ: a clinical study of kidney transplantation', *Psychoanalytic Quarterly*, 42, 364–84.

Bayliss, L. E. (1966) *Living Control Systems*, The English Universities Press, London.

Beckett, B. and Gallagher, R. M. (1989) *Coordinated Science: Biology*, Oxford University Press, Oxford.

Bell, S. E. (1994) 'Translating science to the people: Updating *The New Our Bodies, Ourselves*', *Women's Studies International Forum*, 17, 9–18.

Bendelow, G. and Williams, S. J. (1995) 'Transcending the dualisms? Towards a sociology of pain', *Sociology of Health and Illness*, 17 (2), 139–65.

Benison, S., Clifford Barger, A. and Wolfe, E. L. (1987) *Walter B. Cannon: the Life and Times of a Young Scientist*, Harvard University Press, Cambridge, Massachusetts.

Benison, S., Barger, A. C. and Wolfe, E. L. (1991) 'Walter B. Cannon and the mystery of shock: a study of Anglo-American cooperation in World War I', *Medical History*, 35, 217–49.

REFERENCES

Bennett, P. (1993) 'Critical clitoridectomy: female sexual imagery and feminist psychoanalytic theory', *Signs*, 18, 235–59.

Benton, T. (1991) 'Biology and social science: why the return of the repressed should be given a (cautious) welcome', *Sociology*, 25, 1–29.

Bernard, C. (1865) 'An introduction to the study of experimental medicine'. Reprinted in: Langley, L. L. (ed.), (1973) *Homeostasis: Origins of the Concept*, Dowden, Hutchinson and Ross, Stroudsburg, Pennsylvania.

Berrill, N. J. (1970) *Biology in Action*, Heinemann Educational Books, London.

Billings, S. (1971) 'Concepts of nerve fibre development, 1839–1930', *Journal of the History of Biology*, 4, 275–305.

Biology and Gender Study Group (1989) 'The importance of feminist critique for contemporary cell biology', in N. Tuana (ed.), *Feminism and Science*, Indiana University Press, Bloomington.

Birke, L. (1986) *Women, Feminism and Science: the Feminist Challenge*, Wheatsheaf, Brighton.

Birke, L. (1989) 'How do gender differences in behaviour develop? A reanalysis of the role of early experience', in P. Bateson and P. Klopfer (eds), *Perspectives in Ethology: vol 8 Whither Ethology?*, Plenum Press, New York.

Birke, L. (1992) 'In pursuit of difference', in L. Keller and G. Kirkup (eds), *Inventing Women*, Polity Press, London.

Birke, L. (1994) *Feminism, Animals and Science: the Naming of the Shrew*, Open University Press, Buckingham.

Birke, L. (1995) 'On keeping a respectful distance', in Birke and Hubbard, eds., *Reinventing Biology: Respect for Life and the Creation of Knowledge*, Indiana University Press, Bloomington.

Birke, L., (1997) 'Science and animals – or, why Cyril won't win the Nobel Prize', *Animal Issues*, 1 (1), 45–55.

Birke, L. (1998) 'Biological sciences', in A. Jaggar and I. Young (eds), *A Companion to Feminist Philosophy*, Blackwell, Oxford.

Birke, L. and Hubbard, R. (eds) (1995) *Reinventing Biology: Respect for Life and the Creation of Knowledge*, Indiana University Press, Bloomington.

Birke, L. and Michael, M. (1997) 'Hybrids, rights and their proliferation', *Animal Issues*, 1, 1–19.

Birke, L. and Michael, M. (1998) 'The heart of the matter: animal bodies, ethics and species boundaries', *Society and Animals*, 6, 245–62.

Birke, L. and Smith, J. (1994) 'Animals in experimental reports', *Society and Animals*, 3, 23–42.

Birke, L. I. A., Holzhausen, C., Murphy, S. and Sadler, D. (1984) 'Effects of neonatal medroxyprogesterone acetate on postnatal sexual differentiation of female rats', *Journal of Reproduction and Fertility*, 71, 309–14.

Black, J. E., Sirevaag, A. M., Wallace, C. W., Savin, M. H. and Greenough, W. T. (1989) 'Effects of complex experience on somatic growth and organ development in rats', *Developmental Psychobiology*, 22, 727–52.

REFERENCES

Blackwell, A. B. (1875) 'Sex and evolution'. Reprinted in Rossi, A. (ed.) (1973), *The Feminist Papers*, Bantam, New York.

Bleier, R. (1984) *Science and Gender*, Pergamon, Oxford.

Blume, S. S. (1997) 'The rhetoric and counter-rhetoric of a "bionic" technology', *Science, Technology and Human Values*, 22, 31–56.

Bordo, S. (1993) *Unbearable Weight: Feminism, Western Culture and the Body*, University of California Press, Berkeley.

Borrell, M. (1976) 'Organotherapy, British physiology, and discovery of the internal secretions', *Journal of the History of Biology*, 9, 235–68.

Boston Women's Health Collective (1973) *Our Bodies, Ourselves*, Simon and Schuster, New York.

Bradford Cannon (1975) 'W. B. Cannon: Personal reminiscences', in C. M. Brooks, K. Koizumi and J. O. Pinkston (eds), *The Life and Contributions of Walter Bradford Cannon 1871–1945: His influence in the development of physiology in the Twentieth Century*, State University of New York Press, New York.

Braidotti, R. (1989) 'Organs without bodies', *Differences*, 1, 147–61.

Braidotti, R. (1994) *Nomadic Subjects: Embodiment and Sexual Difference in Contemporary Feminist Theory*, Columbia University Press, New York.

Bray, R. (1994) 'First stirrings', in P. Foster (ed.) *Minding the Body: Women Writers on Body and Soul*, Doubleday, New York.

Brighton Women and Science Group (1980) *Alice Through the Microscope: the Power of Science over Women's Lives*, Virago, London.

British Medical Association (1992) *Medicine Betrayed: the Participation of Doctors in Human Rights Abuses*, Zed Press, London.

Butler, J., (1993) *Bodies that Matter: on the Discursive Limits of "Sex"*, Routledge, London.

Caddick, A. (1992) 'Feminist and postmodern', *Arena*, 99/100, 112–28.

Calnan, M. and Williams, S. (1992) 'Images of scientific medicine', *Sociology of Health and Illness*, 14 (2), 233–54.

Canguilhem, G. (1988) *Ideology and Rationality in the History of the Life Sciences*, Massachusetts Institute of Technology, Cambridge, Massachusetts.

Cannon, W. B. (1932) *The Wisdom of the Body*, W. W. Norton & Company, Inc., New York.

Cartwright, L. (1998) 'A cultural anatomy of the Visible Human Project', in P. Treichler, L. Cartwright and C. Penley (eds), *The Visible Woman: Imaging Technologies, Gender and Science*, New York University Press, New York.

Clarke, A. E. (1990) 'Controversy and the development of reproductive sciences', *Social Problems*, 37, 18–37.

Classen, C. (1993) *Worlds of Sense: Exploring the Senses in History and Across Cultures*, London, Routledge.

Crook, P. (1998) 'Human pugnacity and war: some anticipations of sociobiology 1880–1919', *Biology and Philosophy*, 13, 263–88.

190

REFERENCES

Cross, S. J. and Albury, W. R. (1987) 'Walter B. Cannon, L. J. Henderson and the Organic Analogy', *Osiris*, 3, 165–92.

Davis, K. (ed.), (1997) *Embodied Practices: Feminist Perspectives on the Body*, Sage, London.

Davis, K. (1997a) '"My body is my art": cosmetic surgery as feminist utopia?' in K. Davis (ed.), *Embodied Practices: Feminist Perspectives on the Body*, Sage, London.

Davis, M. (1998) 'Biomedical control and diabetes care', *Science as Culture*, 7, 69–93.

Davis, G. P. and Park, E. (1984) *The Heart: The Living Pump*, Torstar Books, New York.

Dawkins, R. (1976) *The Selfish Gene*, Oxford University Press, Oxford.

De Beauvoir, S. (1969) *The Second Sex*, New English Library, London.

Deleuze, G. and Guattari, F. (1987) *A Thousand Plateaus: Capitalism and Schizophrenia*, University of Minnesota Press, Minneapolis.

Dickinson (1892) 'Diseases of the uterus', in H. A. Hare (ed.), *A System of Practical Therapeutics*, Vol. III, Young J. Pentland, Edinburgh and London.

Diprose, R. (1994) *The Bodies of Women: Ethics, Embodiment and Sexual Difference*, Routledge, London.

Donovan, B. (1988) *Humors, Hormones and the Mind*, Macmillan, London.

Donovan, J. (1990) 'Animal rights and feminist theory', *Signs*, 15, 350–75.

Duden, B. (1991) *The Woman Beneath the Skin: A Doctor's Patients in Eighteenth Century Germany*, Harvard University Press, Cambridge, Massachusetts.

Duden, B. (1993) *Disembodying Women: Perspectives on Pregnancy and the Unborn*, Harvard University Press, Cambridge, Massachusetts.

Dyer, R. (1997) *White*, Routledge, London.

Eagleton, T. (1993) 'It's not quite true that I have a body, and not quite true that I am one either', *London Review of Books*, 27 (5), 7–8.

Eagleton, T. (1996) *The Illusions of Postmodernism*, Blackwell, Oxford.

Evans, G. (1998) 'Man made', *The Guardian*, 2 September; G2, 2–3.

Falk, P. (1994) *The Consuming Body*, Sage, London.

Farrant, W. and Russell, J. (1985) *'Beating Heart Disease': a case study in the production of Health Education Council publications*, Institute of Education, London.

Fausto-Sterling, A. (1992) *Myths of Gender: Biological Theories about Women and Men*, 2nd edition, Basic Books, New York.

Fausto-Sterling, A. (1989) 'Life in the XY corral', *Women's Studies International Forum*, 12, 319–31.

Ferguson, H. (1997) 'Me and my shadows: on the accumulation of body-images in Western society, Part Two – the corporeal forms of modernity', *Body and Society*, 3, 1–31.

Figlio, K. M. (1976) 'The metaphor of organization: an historiographical perspective on the bio-medical sciences of the early nineteenth century', *History of Science*, xiv, 17–53.

REFERENCES

Figlio, K. (1996) 'Knowing, loving and hating nature: a psychoanalytic view', in G. Robertson, M. Mash, L. Tickner, J. Bird, B. Curtis and T. Putnam (eds), *FutureNatural: Nature, Science, Culture*, Routledge, London.

Firestone, S. (1979) *The Dialectic of Sex: The Case for Feminist Revolution*, The Women's Press, London.

Foos, L. (1996) *Ex Utero*, Headline Books, London.

Foucault, M. (1973) *The Birth of the Clinic*, Tavistock, London.

Foucault, M., (1979) *The History of Sexuality, Volume 1: Introduction*, Penguin, London.

Frank, R. G. (1994) 'Instruments, nerve action and the all-or-none principle', *Osiris*, 9, 208–35.

Frank, A. (1996) 'Reconciliatory alchemy: bodies, narratives and power', *Body and Society* 2, 53–71.

Fuss, D. (1989) *Essentially Speaking: Feminism, Nature and Difference*, Routledge, London.

Gaard, G. (1993) 'Living interconnections with animals and nature', in G. Gaard (ed.), *Ecofeminism: Women, Animals, Nature*, Temple University Press, Philadelphia.

Ganong, W. F. (1973) *Review of Medical Physiology*, sixth edition, Lange Medical Publications, Los Altos, California.

Garrety, K. (1997) 'Social worlds, actor-networks and controversy: the case of cholesterol, dietary fat and heart disease', *Social Studies of Science*, 27, 727–73.

Gatens, M. (1996) *Imaginary Bodies: Ethics, Power and Corporeality*, Routledge, London.

Gilman, S. L. (1992) 'Black bodies, white bodies: towards an iconography of female sexuality in late nineteenth century art, medicine and literature', in J. Donald and A. Rattansi (eds), *'Race', Culture and Difference*, Sage, London.

Good, B. (1997) *Medicine, Rationality and Experience: An Anthropological Perspective*, Cambridge University Press, Cambridge.

Goodwin, B. (1994) *How the Leopard Changed its Spots*, Phoenix, London.

Goodwin, B. (1996) 'The evolution of cooperative systems', in P. Bunyard (ed.), *Gaia in Action: Science of the Living Earth*, Floris Books, Edinburgh.

Gross, A. G. (1990) *The Rhetoric of Science*, Harvard University Press, Cambridge, Massachusetts.

Gross, M. B. and Averill, M. B. (1983) 'Evolution and patriarchal myths of scarcity and competition', in S. Harding and M. Hintikka (eds), *Discovering Reality: Feminist Perspectives on Epistemology, Metaphysics, Methodology and Philosophy in Science*, Reidel, London.

Grosz, E. (1994) *Volatile Bodies: Toward a Corporeal Feminism*, Indiana University Press, Bloomington.

Grosz, E. (1995) *Space, Time and Perversion*, Routledge, London.

REFERENCES

Grüsser, O.-J. (1994) 'On the history of the ideas of efference copy and reafference', in C. Debru (ed.), *Essays in the History of the Physiological Sciences*, Rodopi, Amsterdam.

Gudding, G. (1996) 'The phenotype/genotype distinction', *Journal of the History of Ideas*, 57, 525–45.

Guerrini, A. (1989) 'The ethics of animal experimentation in seventeenth-century England', *Journal of the History of Ideas*, 50, 391–408.

Hall, D. L. (1976) 'The critic and the advocate: contrasting British views on the state of endocrinology in the early 1920s', *Journal of the History of Biology*, 9, 269–85.

Halpin, Z. T. (1989) 'Scientific objectivity and the concept of "the other"', *Women's Studies International Forum*, 12, 285–94.

Hamer, D. H., Hu, S., Magnuson, V. L., Hu, N. and Pattatucci, A. M. L. (1993) 'A linkage between DNA markers on the X chromosome and male sexual orientation', *Science*, 261, 321–27.

Haraway, D. (1989) *Primate Visions: Gender, Race and Nature in the World of Modern Science*, Routledge, London.

Haraway, D. (1991) *Simians, Cyborgs and Women*, Free Association, London.

Haraway, D. (1991a) 'A cyborg Manfesto: science, technology and socialist-feminism in the late twentieth century', in D. Haraway, *Simians, Cyborgs and Women*, Free Association, London.

Haraway, D. (1991b) 'The biopolitics of postmodern bodies: constitutions of self in immune system discourse', in D. Haraway, *Simians, Cyborgs and Women*, Free Association, London.

Haraway, D. (1991c) 'Situated knowledges: the science question in feminism and the privilege of partial perspective', in D. Haraway, *Simians, Cyborgs and Women*, Free Association, London.

Haraway, D. (1997) *Modest_Witness@Second_Millennium.FemaleMan_ Meets_ OncoMouse*, Routledge, London.

Harding, S. (1991) *Whose Science? Whose Knowledge? Thinking from Women's Lives*, Open University Press, Buckingham.

Hassenstein, B. (1971) *Information and Control in the Living Organism*, Chapman and Hall, London.

Hekman, S. (1992) *Gender and Knowledge: Elements of a Postmodern Feminism*, Northeastern University Press, Boston.

Hirschauer, S. (1991) 'The manufacture of bodies in surgery', *Social Studies of Science*, 21, 279–319.

Hodgkin, A. (1992) *Chance and Design: Reminiscences of Science in Peace and War*, Cambridge University Press, Cambridge.

Hodgkin, A. and Huxley, A. F. (1952) 'Measurement of current–voltage relations in the membrane of the giant axon of *Loligo*', *Journal of Physiology*, 116, 424–48.

Horigan, S. (1990) *Nature and Culture in Western Discourses*, Routledge, London.

REFERENCES

Houser, R., Konstam, V. and Konstam, M. (1992) 'Transplantation: implications of the heart transplantation process for rehabilitation counsellors', *Journal of Applied Rehabilitation Counseling*, 23, 38–43.

Hubbard, R. (1990) *The Politics of Women's Biology*, Rutgers University Press, New Brunswick and London.

Hubbard, R. and Wald, E. (1993) *Exploding the Gene Myth*, Beacon Press, Boston.

Hubbard, R., Henifin, M. S. and Fried, B. (eds), (1982) *Biological Woman – The Convenient Myth*, Schenkman, Boston.

Ingold, T. (1990) 'An anthropologist looks at biology', *Man*, 25, 208–29.

Irigaray, L. (1985) *This Sex Which is Not One*, Cornell University Press, New York.

Jeffreys, D. (1996) 'Have these patients inherited the donors' characteristics?', *Daily Mail*, 4 June, 51.

Joralemon, D. (1995) 'Organ Wars: the battle for body parts', *Medical Anthropology Quarterly*, 9, 335–56.

Jordanova, L. (1991) *Sexual Visions*, Harvester, Brighton.

Kantrowitz, A. (1996) *A Knife to the Heart*, Part 2, BBC Television, 14 May.

Kauffman, S. (1995) *At Home in the Universe: The Search for the Laws of Complexity*, Penguin, Harmondsworth.

Kay, L. (1997) 'Cybernetics, information, life: the emergence of scriptural representations of heredity', *Configurations* 5, 23–91.

Keller, E. F. (1992), *Secrets of Life, Secrets of Death: Essays on Language, Gender and Science*, Routledge, London.

Keller, E. F. (1995) *Refiguring Life: Metaphors of Twentieth-Century Biology*, University of Columbia Press, New York.

Keller, E. F. (1996) 'The biological gaze', in G. Robertson, M. Mash, L. Tickner, J. Bird, B. Curtis and T. Putnam (eds), *FutureNatural: Nature/Science/Culture*, Routledge, London.

Keller, E. F. and Grontkowski, C. R. (1983) 'The mind's eye', in S. Harding and M. Hintikka (eds), *Discovering Reality: Feminist Perspectives on Epistemology, Metaphysics, Methodology and Philosophy of Science*, Reidel, Dordrecht.

Kevles, D. and Geison, G. L. (1995) 'The experimental life sciences in the twentieth century', *Osiris* 10, 97–121.

Kimbrell, A. (1993) *The Human Body Shop: the Engineering and Marketing of Life*, London, Harper Collins.

Klinge, I. (1997) 'Female bodies and brittle bones: medical interventions in osteoporosis', in K. Davis (ed.), *Embodied Practices: Feminist Perspectives on the Body*, Sage, London.

Knorr-Cetina, K. D. (1983) 'The ethnographic study of scientific work: towards a constructivist interpretation of science', in K. D. Knorr-Cetina and M. Mulkay (eds), *Science Observed: Perspectives on the Social Study of Science*, Sage, London.

REFERENCES

Kremer, R. L. (1990) *The Thermodynamics of Life and Experimental Physiology 1779–1880*, Garland Publishing Inc., New York and London.

Lamb, D. (1990) *Organ Transplants and Ethics*, Routledge, London.

Laqueur, T. (1990) *Making Sex: Body and Gender from the Greeks to Freud*, Harvard University Press, Cambridge, Massachusetts.

Latour, B. (1983) 'Give me a laboratory and I will raise the world', in K. Knorr-Cetina and M. Mulkay (eds), *Science Observed*, Sage, London.

Latour, B. (1987) *Science in Action*, Buckingham, Open University Press.

Latour, B. (1993) *We Have Never Been Modern*, Harvester, Hemel Hempstead.

LeVay, S. (1991) 'A difference in hypothalamic structure between hetero-sexual and homosexual men', *Science*, 253, 1034–7.

LeVay, S. (1993) *The Sexual Brain*, MIT Press, Cambridge, Massachusetts.

Lock, M. (1993) *Encounters with Aging: Mythologies of Menopause in Japan and North America*, University of California Press, Berkeley.

Lock, M. (1997) 'Decentering the natural body: making difference matter', *Configurations*, 5, 267–92.

Longino, H. (1990) *Science as Social Knowledge: Values and Objectivity in Scientific Inquiry*, Princeton University Press, New Jersey.

Lupton, D. (1994) *Medicine as Culture: Illness, Disease and the Body in Western Societies*, Sage, London.

Lupton, D. (1996) *Food, the Body and the Self*, Sage, London.

Lykke, N. (1996) 'Between monsters, goddesses and cyborgs: feminist confrontations with science', in N. Lykke and R. Braidotti (eds), *Between Monsters, Goddesses and Cyborgs: Feminist Confrontations with Science, Medicine and Cyberspace*, Zed Press, London.

Lynch, M. (1990) 'The externalized retina: selection and mathematization in the visual documentation of objects in the life sciences', in M. Lynch and S. Woolgar (eds), *Representation in Scientific Practice*, MIT Press, Cambridge, Massachusetts.

Lyon, M. (1997) 'The material body, social processes and emotion: "techniques of the body" revisited', *Body and Society*, 3, 83–101.

Lyster, R. A. (1966) *A First Course in Hygiene*, University Tutorial Press, London.

Mackean, D. G. (1965) *Introduction to Biology*, John Murray, London.

Margulis, L., Guerrero, R. and Bunyard, P. (1996) 'We are all symbionts', in P. Bunyard (ed.), *Gaia in Action: Science of the Living Earth*, Floris Books, Edinburgh.

Marshall, L. H., (1983) 'The fecundity of aggregates: the axonologists at Washington University, 1922–1942', *Perspectives in Biology and Medicine*, 26, 613–36.

Marshall, H. (1996) 'Our bodies ourselves: why we should add old fashioned empirical phenomenology to the new theories of the body', *Women's Studies International Forum*, 19, 253–65.

REFERENCES

Martin, E. (1989) 'The Woman in the Body, Open University Press, Milton Keynes.

Martin, E., (1994) Flexible Bodies: Tracking Immunity in American Culture from the days of Polio to the Age of AIDS, Beacon Press, Boston.

Martin, E. (1996) 'Citadels, rhizomes and string figures', in A. Aranowitz, B. Martinsons and M. Menser (eds), Technoscience and Cyberculture, Routledge, London.

Marvin, C. (1992) When Old Technologies Were New: Thinking about Electric Communication in the Late Nineteenth Century, Oxford University Press, Oxford.

Masters, J. (1995) 'rEvolutionary theory: reinventing our origin myths', in L. Birke and R. Hubbard (eds), Reinventing Biology, Indiana University Press, Bloomington.

McFadden, D. and Pasanen, E. (1998) 'Comparison of the auditory systems of heterosexuals and homosexuals: click-evoked otoacoustic emissions', Proceedings of the National Academy of Sciences, 95, 2709–13.

Merchant, C. (1982) The Death of Nature: Women, Ecology and the Scientific Revolution, Wildwood, London.

Messing, K. and Mergler, D. (1995) '"The rat couldn't speak but we can": inhumanity in occupational health research', in Birke, L. and Hubbard, R. (eds), Reinventing Biology, Indiana University Press, Bloomington.

Midgley, M. (1992) Science as Salvation: A Modern Myth and its Meaning, Routledge, London.

Mitchell, L. M. and Georges, E. (1997) 'Cross-cultural cyborgs: Greek and Canadian women's discourse on fetal ultrasound', Feminist Studies, 23, 373–401.

Mohacsi, P. J., Blumer, C. E., Quine, S. and Thompson, J. F. (1995) 'Aversion to xenotransplantation', Nature, 378, 434.

Moir, A. and Moir, B. (1998) Why Men Don't Iron: the Real Science of Gender Studies, Harper Collins, London.

Moore, C. L. (1982) 'Maternal behavior is affected by hormonal condition of pups', Journal of Comparative Physiological Psychology, 96, 123–9.

Moore, L. J. and Clarke, A. E. (1995) 'Clitoral conventions and transgressions: graphic representations in anatomy texts, c1900–1991', Feminist Studies, 21, 255–301.

Morrison, T. (1993) Playing in the Dark: Whiteness and the Literary Imagination, Vintage Books, New York.

Morse, M. (1994) 'What do cyborgs eat? Oral logic in an information society', in G. Bender and T. Druckrey (eds), Culture on the Brink: Ideologies of Technology, Bay Press, Seattle.

Myers, G. (1990) Writing Biology: Texts in the Social Construction of Scientific Knowledge, University of Wisconsin Press, Madison.

Myers, G., (1990a) 'Every picture tells a story: Illustrations in E. O. Wilson's Sociobiology', in M. Lynch and S. Woolgar (eds), Representation in Scientific Practice, MIT Press, Cambridge, Massachusetts.

REFERENCES

Newman, S. (1995) 'Carnal boundaries: the commingling of flesh in theory and practice', in L. Birke and R. Hubbard (eds) *Reinventing Biology*, Indiana University Press, Bloomington.

Nuffield Council for Bioethics (1996) *Animal to Human Transplants: the Ethics of Xenotransplantation*, Report.

Okeley, J. (1986) *Simone de Beauvoir*, Virago, London.

Orbach, S. (1979) *Fat is a Feminist Issue*, Hamlyn, Feltham, Middlesex.

Oudshoorn, N. (1993) 'United we stand: the pharmaceutical industry, laboratory and clinic in the development of sex hormones into scientific drugs, 1920–1940', *Science, Technology and Human Values*, 18, 5–24.

Oudshoorn, N. (1994) *Beyond the Natural Body: An Archeology of Sex Hormones*, Routledge, London.

Oyama, S. (1985) *The Ontogeny of Information*, Cambridge University Press, Cambridge.

PAHO/WHO (1997) *A Workshop on Gender, Health and Development: a Facilitators Guide*, PAHO/WHO, Washington, DC.

Park, K. (1997) 'The rediscovery of the clitoris: French Medicine and the Tribade, 1570–1620', in D. Hillman and C. Mazzio (eds), *The Body in Parts: Fantasies of Corporeality in Early Modern Europe*, Routledge, London.

Parmley, W. W. (1979) 'Circulatory function and control', in P. B. Beeson, W. McDermott, and J. B. Wyngaarden (eds), *Cecil Textbook of Medicine*, W. B. Saunders Company, Philadelphia.

Pask, G. (1966) 'Comments on the cybernetics of ethical, sociological and psychological systems', in N. Wiener and J. P. Schade (eds), *Progress in Biocybernetics*, Vol. 3, Elsevier, Amsterdam.

Pennisi, E. (1996) 'Teetering on the brink of danger', *Science*, 271, 1665–7.

Petchesky, R. (1987) 'Fetal images: the power of visual culture in the politics of reproduction', *Feminist Studies*, 13, 263–92.

Petherbridge, D. and Jordanova, L. (1997) *The Quick and the Dead: Artists and Anatomy*, National Touring Exhibitions, The South Bank Centre, London.

Plumwood, V. (1993) *Feminism and the Mastery of Nature*, Routledge, London.

Pouchelle, M.-C. (1990) *The Body and Surgery in the Middle Ages*, Rutgers University Press, New Brunswick, New Jersey.

Radley, A. (1995) 'The elusory body and social constructionist theory', *Body and Society*, 1, 3–23.

Radley, A. (1996) 'The critical moment: time, information and medical expertise in the experience of patients receiving coronary bypass surgery', in S. J. Williams and M. Calnan (eds), *Modern Medicine: Lay Perspectives and Experiences*, UCL Press, London.

Reiser, S. J. (1978) *Medicine and the Reign of Technology*, Cambridge University Press, Cambridge.

Rich, A. (1976) *Of Woman Born: Motherhood as Experience and Institution*, Virago, London.

Rich, A. (1986) 'Contradictions: tracking poems, part 18', in L. Anderson (ed.) (1991) *Sisters of the Earth: Women's Prose and Poetry about Nature*, Vintage Books, New York.

Richardson, R. (1988) *Death, Dissection and the Destitute*, Pelican, London.

Rivera Fuentes, C. (1997) 'Two stories, three lovers and the creation of meaning in a Black Lesbian autobiography: a diary', in H. S. Mirza (ed.), *Black British Feminisms*, Routledge, London.

Romanyshyn, R. D. (1989) *Technology as Symptom and Dream*, Routledge, London.

Romer, A. S. (1970) *The Vertebrate Body*, W. B. Saunders, London.

Rose, H. (1994) *Love, Power and Knowledge: Towards a Feminist Transformation of the Sciences*, Polity Press, London.

Rose, H. (1995) 'Learning from the new priesthood and the shrieking sisterhood: debating the life sciences in Victorian England', in L. Birke and R. Hubbard (eds), *Reinventing Biology*, Indiana University Press, Bloomington

Rose, H. (1997) 'Goodbye truth, hello trust: prospects for feminist science and technology studies at the millennium?', in M. Maynard (ed.), *Science and the Construction of Women*, UCL Press, London.

Rose, S. (1997a) *Lifelines: Biology, Freedom, Determinism*, Penguin, Harmondsworth, Middlesex.

Rosenberg, R. (1982) *Beyond Separate Spheres: Intellectual Roots of Modern Feminism*, Yale University Press, New Haven.

Rosser, S. V. (1992) *Biology and Feminism*, Twayne Publishers, New York.

Rossi, A. (1973) *The Feminist Papers*, Bantam, New York.

Sauvy, A. (1989) *Le Miroir Du Coeur: Quatre Siecles d'images savantes et populaires*, CERF, Paris.

Scarry, E. (1985) *The Body in Pain*, Oxford University Press, Oxford.

Scarry, E. (1995) 'The merging of bodies and artifacts in the social contract', in G. Bender and T. Druckrey (eds), *Culture on the Brink: Ideologies of Technology*, Bay Press, Seattle.

Schiebinger, L. (1989) *The Mind Has No Sex? Women in the Origins of Modern Science*, Harvard University Press, Cambridge, Massachusetts.

Schiebinger, L. (1993) *Nature's Body: Sexual Politics and the Making of Modern Science*, Beacon Press, Boston.

Schmidt-Nielsen, K. (1972) *How Animals Work*, Cambridge University Press, Cambridge.

Schulster, D., Burstein, S. and Cooke, B. A. (1976) *Molecular Endocrinology of the Steroid Hormones*, John Wiley and Sons, London.

Seltzer, M. (1992) *Bodies and Machines*, Routledge, London.

Selzer, R. (1974) *Mortal Lessons: Notes on the Art of Surgery*, Simon and Schuster, New York.

Shapin, S. (1996) *The Scientific Revolution*, University of Chicago Press, Chicago.

Shapin, S. and Shaffer, S. (1985) *Leviathan and the Air-pump: Hobbes, Boyle*

REFERENCES

and the Experimental Life, Princeton University Press, Princeton.

Sharp, L. (1995) 'Organ transplantation as a transformative experience: anthropological insights into the restructuring of the self', *Medical Anthropology Quarterly*, 9, 357–89.

Shildrick, M. (1997) *Leaky Bodies and Boundaries: Feminism, Postmodernism and (Bio)ethics*, Routledge, London.

Shildrick, M. and Price, J. (1998) *Vital Signs: Feminist Reconfigurations of the Bio/logical Body*, Edinburgh University Press, Edinburgh.

Shilling, C. (1993) *The Body and Social Theory*, Sage, London.

Shilling, C. and Mellor, P. A. (1996) 'Embodiment, structuration theory and modernity: mind/body dualism and the repression of sensuality', *Body and Society*, 2, 1–15.

Shiva, V. (1995) 'Democratizing biology: reinventing biology from a feminist, ecological and Third World perspective', in L. Birke and R. Hubbard (eds), *Reinventing Biology*, Indiana University Press, Bloomington.

Shodhini (1997) *Touch Me, Touch me Not: Woman Plants and Healing*, Kali for Women, New Delhi.

Simmons, F. N. (1997) 'My body, myself: how does a Black woman do sociology?', in H. S. Mirza (ed.), *Black British Feminisms*, Routledge, London.

Singer, C. (1957) *A Short History of Anatomy from the Greeks to Harvey*, Dover Publications, New York.

Skeggs, B. (1997) *Formations of Class and Gender: Becoming Respectable*, Sage, London.

Spanier, B. (1995) *Im/Partial Science: Gender Ideology in Molecular Biology*, Indiana University Press, Bloomington.

Spelman, E. (1988) *Inessential Woman: Problems of Exclusion in Feminist Thought*, Beacon Press, Boston.

Stabile, C. (1994) *Feminism and the Technological Fix*, Manchester University Press, Manchester.

Stacey, J. (1997) *Teratologies: A Cultural Study of Cancer*, Routledge, London.

Stacey, J. (1997a) 'Feminist Theory: capital F, capital T', in V. Robinson and D. Richardson (eds), *Introducing Women's Studies*, Macmillan, Basingstoke.

Stevens, S. M. (1997) 'Sacred heart and secular brain', in D. Hillman and C. Mazzio (eds), *The Body in Parts: Fantasies of Corporeality in Early Modern Europe*, Routledge, London.

Synnott, A. (1993) *The Body Social: Symbolism, Self and Society*, Routledge, London.

Tauber, A. I. (1994) *The Immune Self: Theory or Metaphor?*, Cambridge University Press, Cambridge.

Taylor, G. R. (1963) *The Science of Life*, Thames and Hudson, London.

Terry, J. (1995) 'Anxious slippages between "us" and "them": A brief history of the scientific search for homosexual bodies', in J. Terry and J. Urla (eds), *Deviant Bodies*, Indiana University Press, Bloomington.

REFERENCES

Tomas, D. (1996) 'Feedback and cybernetics: reimaging the body in the age of cybernetics', in M. Featherstone and R. Burrows (eds), *Cyberspace/ Cyberbodies/Cyberpunk: Cultures of Technological Embodiment*, Sage, London.

Trumpler, M. (1997) 'Converging images: techniques of intervention and forms of representation of sodium-channel proteins in nerve cell membranes', *Journal of the History of Biology*, 30, 55–89.

Tuana, N. (1993) *The Less Noble Sex: Scientific, Religious and Philosophical Conceptions of Woman's Nature*, Indiana University Press, Bloomington.

Turner, B. S. (1984) *The Body and Society: Explorations in Social Theory*, Blackwell, Oxford.

Turner, B. S. (1992) *Regulating Bodies*, Routledge, London.

Vasseleu, C. (1991) 'A "genethics" that makes sense', in R. Diprose and R. Ferrell (eds), *Cartographies: Poststructuralism and the Mapping of Bodies and Spaces*, Allen and Unwin, Sydney.

Wang, X. (1997) 'The body that puts the mind on trial', *Differences*, 9, 95–128.

Weasel, L. (1997) 'The cell in relation: an ecofeminist revision of cell and molecular biology', *Women's Studies International Forum*, 20, 49–59.

Wendell, S. (1996) *The Rejected Body: Feminist Philosophical Reflections on Disability*, Routledge, London.

Wiener, N. (1948) *Cybernetics: or Control and Communication in the Animal and the Machine* (2nd ed., 1961), MIT Press, Cambridge, Massachusetts.

Wijngaard, M. van den (1997) *Reinventing the Sexes: the Biomedical Construction of Femininity and Masculinity*, Indiana University Press, Bloomington.

Williams, S. and Calnan, M. (1996) *Modern Medicine: Lay Perspectives and Experiences*, UCL Press, London.

Williamson, S. and Nowak, R. (1998) 'The truth about women', *New Scientist*, 1 August, pp. 34–5.

Wittig, M. (1975) *The Lesbian Body*, Peter Owen, London.

Wittig, M. (1992) *The Straight Mind*, Harvester Wheatsheaf, London.

Woodward, K. (1994) 'From virtual cyborgs to biological time bombs: technocriticism and the material body', in G. Bender and T. Druckrey (eds), *Culture on the Brink: Ideologies of Technology*, Bay Press, Seattle.

Yeo, M. (1988) 'Leaves and potatoes', in C. McEwan (ed.), *Naming the Waves: Contemporary Lesbian Poetry*, Virago, London.

Young, I. M. (1984) 'Pregnant embodiment: subjectivity and alienation', *Journal of Medicine and Philosophy*, 9, 45–62.

Young, I. M. (1990) *Throwing Like a Girl and Other Essays in Feminist Philosophy and Social Theory*, Indiana University Press, Bloomington.

Young, S., (1992) 'Review: in praise of miraculous plumbing', *New Scientist*, 22 August, 36.

Index